INTERNATIONAL TECHNOLOGICAL UNIVERSITY
This Book is Donated by:
PROF. WAI-KAI CHEN

Date:

XVII Winter Meeting on

LECTURES ON
THERMODYNAMICS

STATISTICAL MECHANICS

XVII Winter Meeting on Statistical Physics

LECTURES ON THERMODYNAMICS AND STATISTICAL MECHANICS

5–8 January, 1988
Oaxtepec, Mexico

Editors
Agustín E González
Instituto de Física
Universidad Nacional Autónoma de México
Carmen Varea
Facultad de Química
Universidad Nacional Autónoma de México

World Scientific
Singapore • New Jersey • London • Hong Kong

Published by

World Scientific Publishing Co. Pte. Ltd.
P O Box 128, Farrer Road, Singapore 9128

USA office: World Scientific Publishing Co., Inc.
687 Hartwell Street, Teaneck, NJ 07666, USA

UK office: World Scientific Publishing Co. Pte. Ltd.
73 Lynton Mead, Totteridge, London N20 8DH, England

Library of Congress Cataloging-in-Publication data is available.

LECTURES ON THERMODYNAMICS AND STATISTICAL MECHANICS

Copyright © 1988 by World Scientific Publishing Co. Pte. Ltd.

All rights reserved. This book, or parts thereof, may not be reproduced in any form or by any means, electronic or mechanical, including photocopying, recording or any information storage and retrieval system now known or to be invented, without written permission from the Publisher.

ISBN 9971-50-718-8

Printed in Singapore by JBW Printers & Binders Pte. Ltd.

FOREWORD

These proceedings contain the lectures presented at the XVII Winter Meeting on Statistical Physics, held in Oaxtepec, Mexico, from the 5th to the 8th of January, 1988.

As a well established scientific tradition in our country, these meetings provide an updated glance to recent advances in a variety of topics in statistical physics. The subject themes are selected in accordance to their novelty and growth worldwide. This year, a special emphasis was put into rapidly growing fields such as polymer statistics, dynamics of colloidal particles, diffusion controlled reactions and universalities on disordered conductors.

The meeting was made possible by the generous support of the Universidad Autónoma Metropolitana; the Instituto de Investigaciones en Materiales, the Instituto de Física, the Facultad de Química and the Facultad de Ciencias, from the Universidad Nacional Autónoma de México. We thank all those institutions for their help

                              Agustín E. González   Carmen Varea

CONTENTS

Foreword v

Brownian Motion in Concentrated Colloidal Dispersions 1
    M. Medina-Noyola

Longtime, Nonpreaveraged Molecular Diffusivity, Sedimentation
Velocity and Taylor Dispersivity of a Fluctuating Cluster of
Interacting Brownian Particles in a Viscous Flow 14
    H. Brenner, A. Nadim & S. Haber

Simple Models For Diffusion Limited Reactions 29
    F. Leyvraz

Soluble Random-Matrix Model for Dissipative Two-Level Systems 45
    P. Pereyra

Macroscopic Approach to Disordered Conductors 61
    P. A. Mello

Equilibrium Polymerization as a Phase Transition 66
    S. C. Greer

Interaction Between Polymers and Colloidal Particles 74
    P. Pincus

Rigorous and Exact Results for a Model Three-Component
Solution 89
    D. A. Huckaby

Effect of the Counterions on the Electrokinetic Properties and Ion
Adsorption in Charged Microporous Media 96
    W. Olivares, M. Huerta, P. Colmenares & J. C. Villegas

Anisotropic Structure of a Simple Liquid 109
    H. J. M. Hanley

Transient Pattern Formation in Nonequilibrium Fluids 125
    R. C. Desai & K. R. Elder

CONTENTS

Foreword v

Brownian Motion in Concentrated Colloidal Dispersions 1
  M. Medina-Noyola

Long-time Nonpreaveraged Molecular Diffusivity, Sedimentation
Velocity and Taylor Dispersivity of a Flocculating Cluster of
Interacting Brownian Particles in a Viscous Flow 16
  H. Brenner, A. Nadim & S. Haber

Simple Models For Diffusion Limited Reactions 39
  G. Zumofen

Soluble Random Matrix Model for Dissipative Two-Level Systems 45
  J.V. Jose

Macroscopic Approach to Disordered Conductors 51
  P.A. Mello

Equilibrium Polymerization as a Phase Transition 60
  S.C. Greer

Interaction Between Polymers and Colloidal Particles 71
  A. Parola

Rigorous and Exact Results for a Model Three-Component
Solution 80
  D.A. Huckaby

Effect of the Counterions on the Electrokinetic Properties and Ion
Adsorption in Charged Microporous Media 96
  F. Olivares, M. Rivera, F. Guzmán & J.C. Hegseth

Anisotropic Structure of a Simple Liquid 100
  W.A.M. Morgado

Transient Pattern Formation in Nonequilibrium Fluids 125
  R.C. Desai & A.R. Bhat

vii

BROWNIAN MOTION IN CONCENTRATED COLLOIDAL DISPERSIONS*

M. Medina-Noyola

Departamento de Física,
Centro de Investigación y de Estudios Avanzados del
Instituto Politécnico Nacional, apartado postal 14-740,
07000 México D.F. MEXICO

ABSTRACT. A theory is presented in which the intermediate- and the long-time tracer-diffusion properties of concentrated colloidal suspensions (with strong hydrodynamic interactions) are written in terms of the static properties of the suspension, and of the short-time tracer-diffusion coefficient. This theory is applied to the interpretation of the recent experimental measurements of the long-time self-diffusion coefficient D_S^L in concentrated hard-sphere suspensions. The theoretical predictions, which involve no adjustable parameters, are found to be in excellent agreement with the experimental results for D_S^L.

1. INTRODUCTION

During the last decade a great progress has been achieved in the understanding of the static and dynamic properties of colloidal dispersions. This is due to the introduction of modern experimental techniques, such as dynamic light scattering, and to the consequent theoretical developements.[1-3] Still, the departure from certain limiting conditions (e.g. infinite dilution, absence of hydrodynamic interactions, etc.,) seems to call for increasingly more sophisticated formal and conceptual tools for its fundamental description. This situation is quite understandable, given the fact that the dynamics of concentrated colloidal dispersions reflects the coupling of two rather complex phenomena. One of them is the effect of the collisions (or direct interactions) between colloidal particles. This is the analog, at the Brownian level, of the effect of collisions between atoms in a simple liquid. The other, however, is a typical feature of Brownian systems, and refers to the hydrodynamic interactions between the particles, mediated by the supporting solvent. The description of these phenomena provided by the many-body Langevin, Fokker-Planck, or Smoluchowsky equations,[1] is conventionally regarded as the fundamental level of description, to which any theory should be referred. Unfortunately, deriving from this level quantitative predictions applicable to concentrated dispersions seems to be a highly nontrivial problem. Therefore, from this point of view, the apparent consensus that these two effects intermingle in a particularly complicated fashion is indeed well-founded.

On the other hand, it is also reasonable to expect that some form of simplification might emerge if this problem could be viewed from a less detailed perspective. With this expectation in mind, we have recently proposed a theoretical approach to describe the effects of collisions between colloidal particles on their tracer-diffusion properties.[4] In this approach, the Brownian motion of a tracer particle is regarded as the contraction of a stochastic description involving only the collective variables (i.e., the local concentration and the local current) of the other diffusing particles, and not their detailed microscopic configuration (*i.e.*, their positions and velocities). Departing at the outset from the conventional point of view, such a theory is based on general principles of the linear irreversible thermodynamic theory of fluctuations, and on the introduction of well-defined approximations. It has been shown that this approach provides a unifying framework for theoretical results concerning the friction on a charged tracer due to its direct interactions with other colloidal particles[4] and with its own ionic atmosphere.[5]

In such previous applications, the effects of hydrodynamic interactions were neglected as a simplifying assumption. The purpose of this paper is to show that the essential ingredients of that theory are general enough to allow for a simple extension applicable to concentrated systems in which hydrodynamic interactions constitute a most important feature. As a result of the present theory, expressions are derived for the intermediate and long-time tracer-diffusion properties, in terms of static quantities (pair potentials and radial distribution functions) and of the short-time tracer-diffusion coefficient. The latter is considered to be an input of the theory, already containing the most essential effects of the hydrodynamic interactions, and will be assumed to be known, either from experimental measurements[6,7], or from theoretical calculations[8-10]. Thus, for given pair potentials describing the direct interactions, a number of concrete predictions concerning the relationship between the short-time and the long-time tracer-diffusion coefficients will be derived, and mechanisms for their experimental verification will be suggested. Here we will show that the theoretical predictions concerning the long-time self-diffusion coefficient in concentrated hard-sphere suspensions are in excellent agreement with recent experimental results.

In the next section we discuss the fundamental assumption of this theory. The starting equations are introduced in section 3, and the general results derived from them are then discussed in section 4. In section 5, pertinent approximations are introduced, whose explicit application to hard-sphere suspensions is presented in section 6 and in the final discussion.

2. LONG AND SHORT TIMES

Let us start with some definitions. In tracer-diffusion experiments one observes averaged properties of the Brownian motion of individual tracer particles. Quantities such as the mean squared displacement, $\langle (\Delta r_T(t))^2 \rangle /3$, are measured in the diffusive regime, *i.e.*, for times $t \gg \tau_B = M_T/\varsigma_T$, where τ_B is the relaxation time of the velocity of the tracer, M_T is its mass and ς_T its friction coefficient. Under representative conditions, τ_B may be of the order of 10^{-8} sec. In the diffusive regime, one must also distinguish between short and long times. It is the existence of this time-scale separation what constitutes the fundamental assumption of the present theory. The short-time regime may be defined by the condition that the r.m.s. displacements of all the suspended particles be very small compared with their mean separation distance, $n^{-1/3}$, where n is their average number concentration. In other words, at short times the spatial configuration of the suspended particles has not changed significantly, and each of them undergoes diffusive motion in the essentially static field of its neighbors. The diffusion coefficient describing the Brownian motion of the tracer in this regime is referred to as the short-time tracer-diffusion coefficient, D_T^S. (In this work the superscript S denotes "short-time", as opposed to "long-time", which will be denoted by the superscript L. The subscript T refers to the "tracer" particle, which is assumed in general different from the other diffusing particles, whose properties will be denoted by the subscript D. If the tracer is identical to the other diffusing particles, tracer-diffusion becomes self-diffusion, and the subscript T may be changed to S). The corresponding short-time friction coefficient ς_T^S is defined by means of the Einstein relation, $D_T^S = k_B T / \varsigma_T^S$. In the absence of hydrodynamic interactions, $D_T^S = D_T^o \equiv k_B T / \varsigma_T^o$, where ς_T^o is given by the Stokes expression, $\varsigma_T^o = 6\pi\eta a_T$, with a_T being the radius of the tracer particle and η the viscosity of the pure solvent. In concentrated systems, the hydrodynamic interactions lead to an increased hydrodynamic friction, and hence, to a corresponding decrease in D_T^S. Such a decrease has been measured[6,7] in concentrated hard-sphere suspensions as a function of volume fraction. The rigourous theoretical interpretation of these results requires the solution of a rather involved many-body hydrodynamic problem. Nevertheless, the results of the approximate theories[10] are found to be in good agreement with the available experimental data.

The long-time regime, on the other hand, corresponds to times much longer than $\tau_I = n^{-2/3}/D_D^S$, which is the time required by a macroparticle to diffuse a mean interparticle distance. Under typical conditions, τ_I may be of the order of 10^{-3} sec. In the limit $t/\tau_I \gg 1$, $\langle (\Delta \mathbf{r}_T(t))^2 \rangle = 6D_T^L t$, and this defines the long-time tracer-diffusion coefficient D_T^L. At such long times, the tracer has collided very many times with the other suspended particles, and this produces an additional friction. Thus, D_T^L is always smaller than D_T^S, as it has been observed experimentally.[7] The quantitative theoretical prediction of this additional friction, and in general, of the coupled effects of the hydrodynamic and direct interactions, on the time-dependent events at intermediate and long times continues to be a major goal of the current research in this field. As is now well known, these time-dependent phenomena may be observed in great detail by means, mainly, of dynamic light scattering techniques. Our approach to the description of these phenomena is now outlined.

3. THE BASIC EQUATIONS

Let us consider an equilibrium ensemble of systems, each consisting of a single tracer particle interacting with the other colloidal particles in a monodisperse suspension. Both, the tracer and the other diffusing particles, will be assumed spherical, and interacting by means of hydrodynamic and direct forces. In particular, the direct interaction between the tracer and any of the other diffusing particles will be described by a pair potential $\psi(r)$. We shall describe the state of the system by means of the variables $\mathbf{V}(t)$ and $n'(\mathbf{r}, t)$, which are the velocity of the tracer and the instantaneous local concentration of the other particles at position $\mathbf{r}$, referred to the center of the tracer. The equilibrium-ensemble average of these variables is 0 and $n^{eq}(r) = n g_{TD}(r)$, where $g_{TD}(r)$ is the radial distribution function of the colloidal particles around the tracer, and n is their bulk number concentration. In reference 4, the local macroparticle current $\mathbf{j}(\mathbf{r}, t)$ was included as a state variable, whose explicit consideration was not required in the absence of hydrodynamic interactions. However, even if hydrodynamic interactions are present, one can always eliminate $\mathbf{j}(\mathbf{r}, t)$ from the description, so that the only state variables are $\mathbf{V}(t)$ and $n'(\mathbf{r}, t)$.

From quite general considerations, one can expect that the relaxation of the equilibrium fluctuations of these state variables obeys a linearized system of equations with the following structure

$$\frac{M_T d\mathbf{V}(t)}{dt} = -\varsigma_T^S \mathbf{V}(t) + \mathbf{f}_T(t) + \int d^3 r \, [\nabla \psi(r)] \overline{\delta n'}(\mathbf{r}, t) \tag{1}$$

and

$$\frac{\partial \delta n'(r,t)}{\partial t} = [\nabla n^{\text{eq}}(r)] \cdot \mathbf{V}(t) \\ - \int_0^t dt' \int d^3 r' \, G'(\mathbf{r}, \mathbf{r}'; t-t') \delta n'(\mathbf{r}', t') + f(\mathbf{r}, t) \quad (2)$$

Equation (1) expresses the total force on the tracer as the sum of the direct interaction forces exerted by the other diffusing particles (the third term on the r.h.s.) plus the hydrodynamic forces exerted by the supporting solvent, represented as a dissipative friction term, plus its corresponding δ-correlated stochastic force, $\mathbf{f}_T(t)$.

The direct interaction term represents the collisions of the tracer with the other diffusing particles. It constitutes a non-dissipative, purely mechanical coupling between the velocity of the tracer and the instantaneous configuration of the other particles, which has been written in terms of the deviation $\delta n'(\mathbf{r}, t)$ of the local concentration profile $n'(\mathbf{r}, t)$, from its equilibrium value $n^{\text{eq}}(r)$ [the prime in $n'(\mathbf{r}, t)$ is a reminder that the origin of the coordinate system $\mathbf{r}$ is at the center of the tracer]. It should be stressed that although the concept of "collisions" may provide sometimes a vivid picture of the dynamical events occurring in the suspension, at the level of description that we are employing, the direct interactions couple the velocity of the tracer with the positions of the other particles in the manner expressed in equation (1), *i.e.*, only through the collective variable $n'(\mathbf{r}, t)$. Furthermore, it is not difficult to see[4] that the form of this coupling, as written in equation (1), is exact, and any fundamental derivation of these time-evolution equations is bound to reproduce this term.

On the other hand, the identification of the friction coefficient appearing in equation (1) with the short-time friction coefficient ς_T^S constitutes a central feature of the present theory and, essentially by definition, is also exact, provided the existence of well-separated time-scales, according to our definition of "short" and "long" times. Thus, at short times, equation (1) describes the diffusive motion of the tracer in the static, conservative field of its virtually immobile neighbours. The friction coefficient appearing in that equation defines, via the Einstein relation, the diffusion coefficient which determines this diffusive motion. This, however, has been our definition of the short-time tracer-diffusion coefficient D_T^S.

The local concentration of particles around the tracer, which at short times may be viewed as a static distribution, has in fact its own dynamics, which manifests itself in the intermediate and long-time motion of the tracer. Also on the basis of the existence of linear regression laws, the time-evolution equation for $\delta n'(\mathbf{r}, t)$ is expected to have the general structure of equation (2). The first term of the r.h.s. of that equation is a linearized streaming term, deriving from the fact that the coordinate $\mathbf{r}$ of the local concentration $n'(\mathbf{r}, t)$ has its origin in the center of the tracer, which moves with velocity $\mathbf{V}(t)$. Like the direct-force term of equation (1), this term has a mechanical, non-dissipative nature. In fact, these two terms are related exactly to each other by an antisymmetry relation.[4]

Besides this mechanical process allowing $\delta n'(\mathbf{r}, t)$ to evolve in time, there are, of course, diffusive processes, which would be present even if the tracer were held fixed, but which are affected in general by its motion and by its field of force. The second term on the right-hand side of this equation represents the diffusive relaxation of $\delta n'(\mathbf{r}, t)$, and the last term is the corresponding fluctuating term, related to the generalized diffusion kernel $G'(\mathbf{r}, \mathbf{r}'; t)$ by a fluctuation-dissipation relation. We have written the most general form for these diffusive terms, very much to emphasize our ignorance of their detailed structure, but also because there is no need at this stage to specify them. Eventually, however, it will be precisely in their definition where some phenomenological ansatz will have to be introduced.

4. THE CONTRACTED DESCRIPTION: GENERAL RESULTS

After these comments on the generality of equations (1) and (2), we proceed to eliminate the variable $\delta n'(\mathbf{r}, t)$ from the description provided by equations (1) and (2) by formally solving equation (2) for $\delta n'(\mathbf{r}, t)$, and inserting the solution in equation (1). This procedure is analogous to that employed in our previous applications of the present method,[4] and will not be detailed here. We should mention, however, that being a particular application of a general theorem on the contraction of non-Markov stochastic processes,[11] its validity is independent of the specific definition of $G'(\mathbf{r}, \mathbf{r}'; t)$. The application of this contraction procedure leads to the following Langevin equation for the tracer,

$$M_T \frac{d\mathbf{V}(t)}{dt} = -\varsigma_T^S \mathbf{V}(t) + \mathbf{f}_T(t) - \int_0^t dt' \, \Delta\varsigma(t - t') \mathbf{V}(t') + \mathbf{F}(t) \qquad (3)$$

where the time-dependent friction term derives from its direct interactions with the other particles, and $\mathbf{F}(t)$ is the corresponding (colored) stochastic force. $\mathbf{F}(t)$ may be written as an expression linear in the fluctuating diffusive term $f(\mathbf{r}, t)$ of equation (2), and of the initial concentration profile $\delta n'(\mathbf{r}, t = 0)$, which is treated as a random variable with an equilibrium probability distribution. The stochastic force $\mathbf{F}(t)$ may be shown to satisfy a fluctuation-dissipation relationship with the time-dependent friction function $\Delta\varsigma(t)$. This function is found to be given by

$$\Delta\varsigma(t) = \frac{k_B T}{3} \int d^3 r \int d^3 r' \, [\nabla n^{\mathrm{eq}}(r)] \cdot (\sigma^{-1}\chi'(t))(\mathbf{r}, \mathbf{r}')[\nabla' n^{\mathrm{eq}}(r')] \qquad (4)$$

where the function $(\sigma^{-1}\chi'(t))(\mathbf{r}, \mathbf{r}')$ is the convolution of $\sigma^{-1}(\mathbf{r}, \mathbf{r}')$ with the collective-diffusion propagator $\chi'(\mathbf{r}', \mathbf{r}'; t)$ observed from the reference frame of the tracer [i.e., the solution of equation (2) without the streaming and fluctuating terms, but satisfying the initial condition $\chi'(\mathbf{r}, \mathbf{r}'; 0) = \delta(\mathbf{r} - \mathbf{r}')$]. $\sigma^{-1}(\mathbf{r}, \mathbf{r}')$ is the inverse function of the static correlation function $\sigma(\mathbf{r}, \mathbf{r}') = \langle \delta n(\mathbf{r}, o)\delta n(\mathbf{r}', o)\rangle$, in the sense that their convolution is a Dirac delta function.

If $\Delta\varsigma(t)$ were known, other important tracer-diffusion properties would follow. For example, the velocity autocorrelation function, $C(t) = \langle \mathbf{V}(t) \cdot \mathbf{V}(0)\rangle/3$, is given by

$$\hat{C}(z) = \int_0^\infty dt \, e^{-zt} C(t) = \frac{k_B T / M_T}{z + \varsigma_T^S/M_T + \Delta\hat\varsigma(z)/M_T}, \qquad (5)$$

where the tilde indicates Laplace transform. The long-time tracer-diffusion coefficient is given by the zero-frequency limit of this equation,

$$D_T^L = \hat{C}(0) = \frac{k_B T}{\varsigma_T^S + \Delta\varsigma(0)} \qquad (6)$$

Before introducing the approximations for $\chi'(\mathbf{r}, \mathbf{r}'; t)$ needed to apply the results in equations (3) and (4) to the analysis of intermediate and long-time dynamics, let us notice that at very short times ($t/\tau_I \ll 1$), the collective propagator $\chi'(\mathbf{r}, \mathbf{r}'; t) \simeq \chi'(\mathbf{r}, \mathbf{r}'; 0) = \delta(\mathbf{r} - \mathbf{r}')$. Then, it is not difficult to show, using equations (3) and (4), that the Langevin equation in equation (3), when averaged over all the possible initial concentration profiles

$n'(\mathbf{r},0)$, reads

$$M_T \frac{d\mathbf{V}(t)}{dt} = -\varsigma_T^S \mathbf{V}(t) + \mathbf{f}_T(t) - k_H[\Delta \mathbf{r}(t)]. \tag{7}$$

Here, $[\Delta \mathbf{r}(t)] = \int_0^t dt' \, \mathbf{V}(t')$ is the tracer's displacement, and k_H is the spring constant of the harmonic potential felt by the tracer in the center of the cage formed by its neighbours, distributed according to their equilibrium radial distribution function, *i.e.*,

$$k_H = \Delta\varsigma(0) = -\frac{1}{3}\int d^3 r \, [\nabla \psi(r)] \cdot [\nabla n^{\mathrm{eq}}(r)]. \tag{8}$$

In deriving this equation, use is made of the following exact relation[4] between $n^{\mathrm{eq}}(r)$ and $\sigma(\mathbf{r},\mathbf{r}')$,

$$k_B T \nabla n^{\mathrm{eq}}(r) = -\int d^3 r' \, \sigma(\mathbf{r},\mathbf{r}') \nabla' \psi(r') \tag{9}$$

Thus, according to equation (7), in an average sense the tracer behaves in the short-time regime as a harmonically-bound Brownian particle, in agreement with the well known "cage" model. We should mention that although equation (8) may be derived from the Smoluchowski equation as an exact short-time condition,[12] its validity only depends on the existence of well-separated time-scales, as the derivation above implies.

5. INTERMEDIATE AND LONG TIMES

Let us now consider the results in equations (3) and (4) for arbitrary times. As indicated above, the specification of $G'(\mathbf{r},\mathbf{r}';t)$ [or equivalently, of $\chi'(\mathbf{r},\mathbf{r}';t)$] is needed in order to discuss the intermediate and long-time tracer-diffusion properties.

The theory presented here is based on the idea that tracer-diffusion properties only depend on rather general features of the collective dynamics represented by $\chi'(\mathbf{r},\mathbf{r}';t)$, so that a simple approximation for this quantity, employed as an input in the general expression for $\Delta\varsigma(t)$ above, should lead to an accurate description of tracer-diffusion properties. In fact, let us argue that even a reasonable short-time approximation for the collective propagator should suffice. The rationale behind this expectation is that the first-order effects of interparticle collisions are explicitly described through the mechanical couplings between the variables $\mathbf{V}(t)$ and $n'(\mathbf{r},t)$, which have been treated above in an essentially exact fashion. Thus, although the effects of collisions also enter implicitly through $\chi'(\mathbf{r},\mathbf{r}';t)$ [see equation (4)], they should manifest themselves only as second-order effects on the dynamics of the tracer. It is these effects what we neglect when a short-time approximation for $\chi'(\mathbf{r},\mathbf{r}';t)$ is employed in equation (4).

Before specifying our approximations for the collective propagator, let us introduce a simplifying general approximation. The relaxation kernel $G'(\mathbf{r},\mathbf{r}';t)$, which describes the diffusive relaxation of $\delta n'(\mathbf{r},t)$ as observed from the reference frame of the tracer, is not simply related with the kernel $G(\mathbf{r},\mathbf{r}';t)$ describing collective diffusion from a laboratory-fixed reference frame. In previous work[4] we have discussed a method to approximate $G'(\mathbf{r},\mathbf{r}';t)$ based on the use of a Fick's diffusion law. There we also discussed an alternative method, in which the propagator χ' is directly related with the collective propagator χ of the laboratory-fixed reference frame. In such a method the field of the tracer is ignored in the calculation of $\chi'(\mathbf{r}, \mathbf{r}'; t)$, which may then be written as

$$\chi'(\mathbf{r}, \mathbf{r}'; t) = \frac{1}{(2\pi)^3} \int d^3k \, \exp[-i\mathbf{k} \cdot (\mathbf{r} - \mathbf{r}')] \chi'(k, t). \tag{10}$$

This allows equation (4) to be written as [see equation (4.14) of reference 4]

$$\Delta\varsigma(t) = \frac{k_B T n}{24\pi^3} \int d^3k \, \frac{[k \, h_{TD}(k)]^2}{S(k)} \chi'(k, t) \tag{11}$$

where $S(k)$ is the static structure factor of the diffusing particles (in which, for simplicity, we have dropped the subscript DD), and $h_{TD}(k)$ is the Fourier transform of $g_{TD}(r) - 1$. As it has been argued before,[4] a simple way to relate the description of the collective diffusive processes observed from the reference frame undergoing Brownian motion, with the description from a laboratory-fixed reference frame, is by means of the following factorization

$$\chi'(k, t) = \chi_T(k, t) \chi(k, t) \tag{12}$$

where $\chi_T(k, t) \equiv \langle \exp[-i\mathbf{k} \cdot \Delta\mathbf{r}(t)] \rangle$ is the tracer-diffusion propagator and $\chi(k, t)$ is the Fourier transform of the collective-diffusion propagator observed from a laboratory-fixed reference frame, in the absence of the field of the tracer. The resulting expression for $\Delta\varsigma(t)$ happens to coincide with the mode-mode coupling approximation introduced by Hess and Klein[13] some years ago in a derivation in which hydrodynamic interactions were neglected. However, the arguments given in reference 4 to derive equation (11) and (12) do not involve any similar restriction.

Equations (11) and (12) still require additional specific approximations for the tracer- and collective-diffusion propagators. As argued at the begining of this section, a short-time approximation for $\chi(k, t)$, in which memory effects are neglected, should suffice. In this so-called "mean field approximation", $\chi(k, t)$ is given by[1]

$$\chi(k, t) = \exp\left[-\frac{k^2 H(k) t}{S(k)}\right] \tag{13}$$

where $H(k)$ is a wave-vector dependent quantity, whose large-k limit is the short-time tracer-diffusion coefficient D_D^S of the diffusing particles. In the absence of hydrodynamic interactions $H(k)$ is simply D_D^o. Let us mention, however, that the approximate calculation[10] of $H(k)$ for hard-sphere suspensions suggests that D_D^S is in general the most inportant contribution to $H(k)$ except at small wave-vectors and/or high volume fractions. Thus, let us introduce the following additional approximation,

$$H(k) = D_D^S. \tag{14}$$

As a result, equations (11)–(14) provide an expression for the friction function $\Delta\varsigma(t)$ in terms only of the tracer-diffusion propagator $\chi_T(k, t)$, since all the other quantities involved are considered known. However, it is precisely the tracer-diffusion properties themselves what we set out to calculate. Hence, we have to introduce an additional (closure) relation between $\Delta\varsigma(t)$ and $\chi_T(k, t)$.

The simplest closure consists in approximating $\chi_T(k, t)$ by its short-time expression, $\exp(-k^2 D_T^S t)$. Still another closure relation involves the long, rather than the short-time limit of $\chi_T(k,t)$. These two closures have been considered in a recent application[12] of the present method to the calculation of the self-diffusion properties of suspensions of highly charged colloidal particles, where hydrodynamic interactions may be neglected. In such an application it was found that the second closure leads to more accurate quantitative results when compared with computer simulation results. In this paper we shall consider both approximations, denoted respectively as MMC1 and MMC2, according to which $\chi_T(k,t)$ is to be approximated by

$$\chi_T(k,t) = \exp(-k^2 \mu D_T^S t), \tag{15}$$

where the parameter μ distinguishes between these two particular choices, according to

$$\mu = \begin{cases} 1, & \text{MMC1} \\ D_T^L/D_T^S, & \text{MMC2}. \end{cases} \tag{15'}$$

In the latter case, it is clear that the parameter μ must be evaluated self-consistently, as indicated below.

Equation (15), along with equations (11)–(14), completes the approximate scheme derived on the basis of our general results of section 4 and of the specific simplifying approximations introduced in this section. From these results, the intermediate- and the long-time tracer-diffusion properties may be calculated in terms of the static properties $g_{TD}(r)$ and $S(k)$, and of the short-time diffusion coefficients, D_T^S and D_D^S, of the tracer and of the other diffusing particles, respectively. In particular, the long-time tracer-diffusion coefficient is given in this approximation by

$$\frac{D_T^L}{D_T^S} = \left(1 + \frac{n(D_T^S/D_D^S)}{24\pi^3} \int d^3k \, \frac{h_{TD}^2(k)}{1 + \mu(D_T^S/D_D^S)S(k)}\right)^{-1} \tag{16}$$

Let us point out that in these results the hydrodynamic interactions only enter as a renormalizing effect on the short-time tracer-diffusion coefficients. Otherwise, they do not differ from the corresponding results for systems without hydrodynamic interactions,[4,12] which correspond to the particular case in which the short-time tracer-diffusion coefficients D_T^S and D_D^S are given just by D_T^o and D_D^o, respectively. This is, of course, a consequence of the approximations introduced in this section to render the description of hydrodynamic interactions manageable in a simple way, but our general results in section 4 allow for departures from this simple scheme. However, the specific application to hard-sphere suspensions in the following section suggests that this is already a very usefull approximation.

6. SELF-DIFFUSION IN CONCENTRATED HARD-SPHERE SUSPENSIONS

One of the most stringent tests of the results above concerns its application to concentrated hard-sphere suspensions. In this section we shall discuss the specific predictions which derive from equation (16) for the long-time self-diffusion coefficient D_S in a monodisperse hard-sphere suspension. In this case, $h_{TD}(k) = (S(k) - 1)/n$, and the static structure factor $S(k)$ can be calculated using the Percus-Yevick approximation.[14] Thus, according to equation (16), the ratio D_S^L/D_S^S is a function only of the hard-sphere

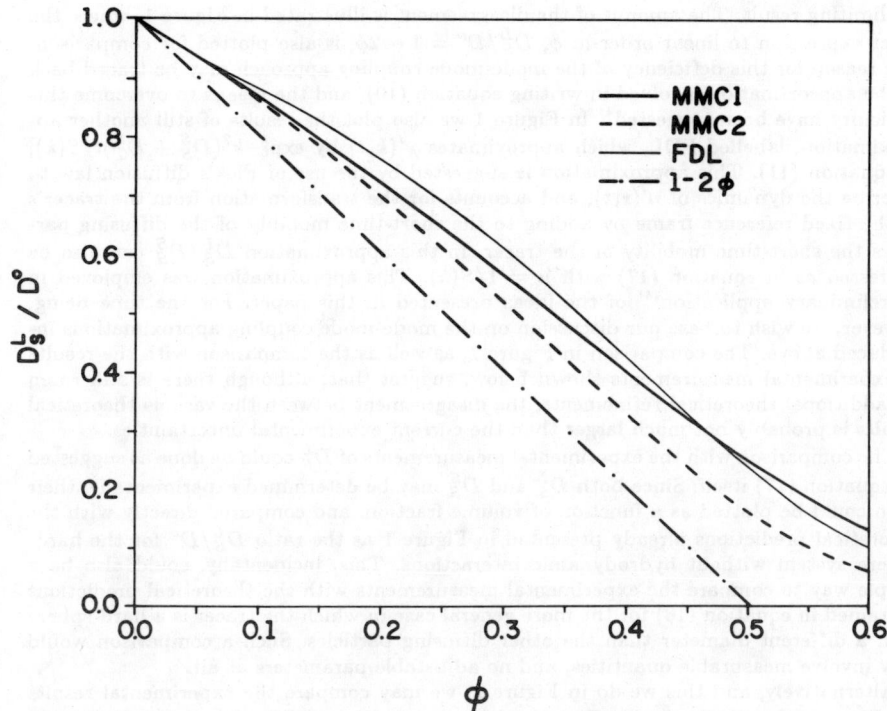

FIGURE 1. Long-time self-diffusion coefficient D_S^L of an idealized (no hydrodynamic interactions) hard-sphere suspension normalized with its short-time self-diffusion coefficient $D_S^S = D^o$, as a function of volume fraction. The calculations correspond to the MMC1, MMC2, and FDL approximations, defined by equation (17) with $\mu = 1$, D_S^L/D_S^S, and $1/S(k)$, respectively, with the Percus-Yevick results for the static structure factor of the hard-sphere system. The straight line is the initial decay, $D_S^L/D^o = 1 - 2\phi$, for a dilute hard-sphere suspension as calculated on the basis of the two-particle Smoluchowski equation.[15, 16]

volume fraction $\phi = \pi n d^3/6$ (with d being the hard-sphere diameter), and is given by

$$\frac{D_S^L}{D_S^S} = \left(1 + \frac{n}{6\pi^2} \int_0^\infty dk \, \frac{[k h(k)]^2}{1 + \mu S(k)}\right)^{-1} \quad (17)$$

where μ is defined in equation (15'). Clearly, in the MMC2 approximation equation (17) must be solved self-consistently for the ratio D_S^L/D_S^S.

Let us notice first that if hydrodynamic interactions are neglected, this result provides an approximation for the ratio D_S^L/D^o of an idealized hard-sphere suspension without hydrodynamic interactions. In Figure 1 we plot the results thus obtained for this ratio within both approximations, MMC1 and MMC2. Although such an idealized system has limited practical relevance, it has been the subject of appreciable theoretical attention,[4, 15–17] particularly since its properties are known exactly in the limit of low concentrations.[15, 16] As it is well-known, the mode-mode coupling approximation fails to reproduce such exact limiting result. The amount of the disagreement is illustrated in Figure 1, where the exact expression to linear order in ϕ, $D_S^L/D^o = 1 - 2\phi$, is also plotted for comparison. The reason for this deficiency of the mode-mode coupling approach may be traced back to the approximation involved in writing equation (10), and the means to overcome this difficulty have been suggested.[) In Figure 1 we also plot the results of still another approximation, labelled FDL, which approximates $\chi'(k,t)$ by $\exp[-k^2(D_S^S + D_T^S)t/S(k)]$ in equation (11). This approximation is suggested by the use of Fick's diffusion law to describe the dynamics of $n'(\mathbf{r},t)$, and accounts for the transformation from the tracer's to the fixed reference frame by adding to the short-time mobility of the diffusing particles the short-time mobility of the tracer. In this approximation D_S^L/D_S^S can also be expressed as in equation (17) with $\mu = 1/S(k)$. This approximation was employed in a preliminary application[18)] of the ideas presented in this paper. For the time being, however, we wish to base our discussion on the mode-mode coupling approximations introduced above. The comparison in Figure 1, as well as the comparison with the results of experimental measurements shown below, suggest that, although there is still room for additional theoretical refinements, the disagreement between the various theoretical results is probably not much larger than the current experimental uncertainties.

The comparison with the experimental measurements of D_S^L could be done as suggested by equation (17) itself. Since both D_S^L and D_S^S may be determined experimentally, their ratio could be plotted as a function of volume fraction, and compared directly with the theoretical predictions already presented in Figure 1 as the ratio D_S^L/D^o for the hard-sphere system without hydrodynamic interactions. This, incidentally, could also be a simple way to compare the experimental measurements with the theoretical predictions contained in equation (16) for the more general case in which the tracer is a hard-sphere with a different diameter than the other diffusing particles. Such a comparison would only involve measurable quantities, and no adjustable parameters at all.

Alternatively, and this we do in Figure 2, we may compare the experimental results for D_S^L, scaled with the Stokes-Einstein diffusion coefficient D^o, with the results of the MMC1 and MMC2 approximations in equation (17), which we may write as

$$\frac{D_S^L}{D^o} = \left(\frac{D_S^L}{D_S^S}\right)\left(\frac{D_S^S}{D^o}\right). \qquad (18)$$

The first factor on the rhs of this equation is given by equation (17). The second factor is to be taken from the experimental measurements of the short-time self-diffusion coefficient. In Figure 2 we plot an empirical fit of the experimental results for D_S^S/D^o reported in Figure 3 of reference 7. We take these values, along with our calculations for the r.h.s. of equation (17) (already displayed in Figure 1), to calculate the predictions of the MMC1 and MMC2 approximations for the long-time self-diffusion coefficient. The results are shown in Figure 2. As can be seen in this figure, there is an excellent qualitative agreement between our theoretical predictions and the experimental results for D_S^L/D^o. Furthermore, considering the fact that our theoretical results involve no fitting parameters at all, we may say that there is also an excellent quantitative agreement with the experimental data, at least at intermediate and large volume fractions. The somewhat poorer agreement at small volume fractions may be due in part to the failure of the MMC approximations to reproduce the exact limiting behaviour in the small concentration regime. The straight line in Figure 2 corresponds to the use of the limiting expression $1 - 2\phi$ for the factor D_S^L/D_S^S in equation (18). Clearly, an approximation which interpolates between this initial decay of D_S at small volume fractions, and the

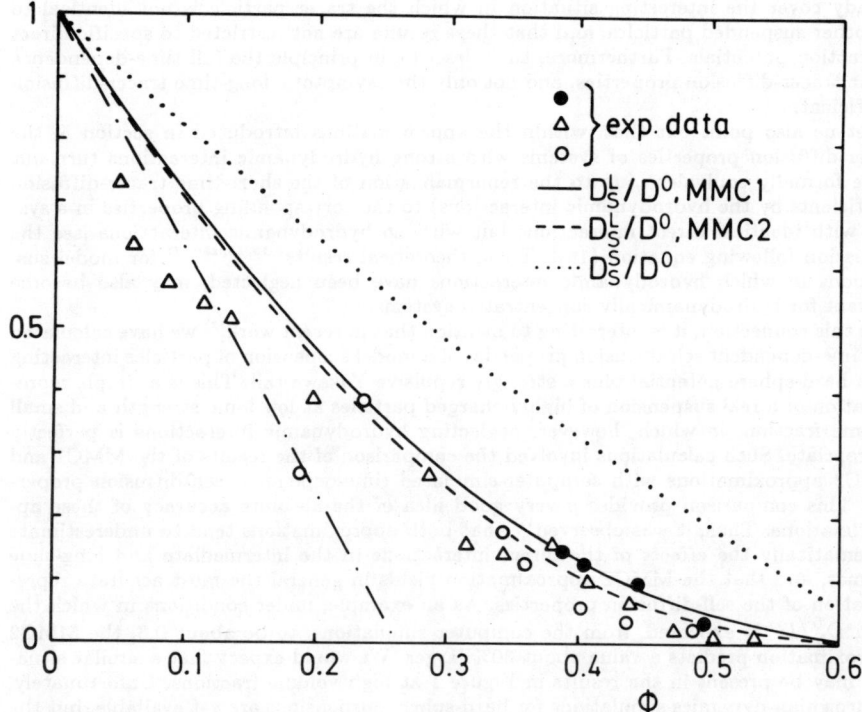

FIGURE 2. Short-time (D_S^S) and long-time (D_S^L) self-diffusion coefficient of a real hard-sphere suspension, normalized with their common infinite-dilution value D^o, as a function of volume fraction. The dotted line is an empirical fit of the experimental measurements of D_S^S reported in Figure 3 of reference 7, and the other symbols represent experimental results for D_S^L as read from Figure 4 of the same reference. The theoretical curves labelled MMC1 and MMC2 are the results obtained from equation (18), with the ratio D_S^L/D_S^S calculated according to equation (17) within the Percus-Yevick approximation. The straight line is $1 - 3.8\phi$.

mode-mode coupling results at higher values of ϕ, would certainly improve further the general agreement between theory and experiment.

7. DISCUSSION

The comparison just presented intends to illustrate the type of predictions which may be derived from the theoretical approach discussed in this paper. Clearly, many other specific conclusions may follow from our general results, and some extensions and improvements of the approximations discussed in this paper could also be thought of. The experimental verification of the results thus produced should involve measurements probably not much more complicated than those employed in the comparison in Figure 2.

Concerning possible extensions, let us mention that the results in section 4 may be easily generalized to describe tracer-diffusion properties in suspensions with several species of Brownian particles, which is the situation most commonly encountered under practical conditions. Similarly, the corresponding extension of the approximate results in section 5 is also straightforward. Let us emphasize, however, that the final results in section 5, already cover the interesting situation in which the tracer particle is not identical to the other suspended particles, and that these results are not restricted to specific direct interaction potentials. Furthermore, they describe in principle the full time-dependence of the tracer-diffusion properties, and not only the asymptotic long-time tracer-diffusion coefficient.

Let us also point out that, within the approximations introduced in section 5, the tracer-diffusion properties of systems with strong hydrodynamic interactions turn out to be formally equivalent (up to the renormalization of the short-time tracer-diffusion coefficients by the hydrodynamic interactions) to the corresponding properties in a system with identical direct interactions but with no hydrodynamic interactions [see the discussion following equation (16)]. Thus, theoretical results[4, 12, 13, 15–17] for model suspensions in which hydrodynamic interactions have been neglected, may also become relevant for hydrodynamically concentrated systems.

In this connection, it is interesting to mention that in recent work,[12] we have calculated the time-dependent self-diffusion properties of a model suspension of particles interacting via a hard-sphere potential plus a strongly repulsive Yukawa tail. This is a simple representation of a real suspension of highly charged particles at low ionic strength and small volume fraction, in which, however, neglecting hydrodynamic interactions is perfectly appropriate. Such calculations involved the comparison of the results of the MMC1 and MMC2 approximations with computer-simulated time-dependent self-diffusion properties. This comparison provides a very good idea of the absolute accuracy of these approximations. Thus, it was observed[12] that both approximations tend to underestimate systematically the effects of the direct interactions in the intermediate and long-time regimes, and that the MMC2 approximation yields in general the most accurate representation of the self-diffusion properties. As an example, under conditions in which the ratio D_S^L/D^o is expected, from the computer simulations, to be about 0.3, the MMC2 approximation predicts a value about 30% larger. We would expect that a similar situation may be present in the results in Figure 1 at high volume fractions. Unfortunately, no Brownian-dynamics simulations for hard-sphere suspensions are yet available, but the comparison with the experimental results for D_S^L in Figure 2 seem to be consistent with this expectation.

Concerning the possible improvements of the specific results for hard-sphere suspensions presented in this paper, let us mention that it is possible to avoid the approximation represented by equation (10), which is responsible for the inaccuracies of the mode-mode coupling results at low volume fractions. An approach based on the general results in section 4 and on the use of Fick's law to describe the dynamics of the particles diffusing around the tracer may be constructed, which does not share this deficiency of the MMC-results.

All these possible extensions and improvements of the theoretical results presented here would be of limited practical use if the pertinent short-time tracer-diffusion coefficients could not be made available by direct experimental measurements. Let us stress, however, the fact that the theory of hydrodynamic interactions has evolved in the recent years[10] to a stage in which reasonably accurate predictions may be derived for these objects. Thus, the results of such theoretical developements could also be employed as the short-time information needed as an input in the application of the ideas presented here.

As a final comment, let us mention that it is not obvious what is the exact relationship between the theoretical results presented in this paper, and the mesoscopic description provided by the many-body Fokker-Planck or Smoluchowsky equations. In other words, it is not clear what is the averaging procedure and the approximations or limits which must be introduced if one attempts to derive from such a level of description the results discussed in this paper. Similarly, it is not clear what is the relationship with the results of the relaxation-effect approach pioneered by Batchelor.[8] In fact, specific results of our theory are found to be inconsistent with results derived with the latter method. We refer specifically to the low concentration result for D_S^L/D^o which, as a consequence of hydrodynamic and direct interactions, is expected to decrease as $1 - 2.1\phi$ at small volume fractions.[8] Instead, the corresponding prediction of our theory is the straight line in Figure 2, which is approximately $1 - 3.8\phi$. Clearly, there is the need to understand the source of this discrepancy. Of course, besides additional theoretical research, the availability of more abundant and accurate experimental information will certainly be of valuable assistance in clarifying this situation. We believe, however, that the ideas presented in this paper, which are based on the firm and general principles of the linear irreversible thermodynamic theory of fluctuations, will constitute at least a useful approximate description of this, otherwise rather involved and complex phenomenon, namely, the coupling between hydrodynamic and direct interactions in concentrated dispersions.

ACKNOWLEDGEMENTS

Some aspects of this work derived from a colaboration with G. Nägele, H. Ruiz-Estrada, A. Vizcarra-Rendón, and R. Klein, whose comments and support are gratefully acknowledged. I also wish to thank Dr. P. Pusey for several interesting remarks concerning the present work. Much of this work developed during a sabbatical leave of the author at Universität Konstanz. It is a pleasure to acknowledge the fine hospitality received from Prof. R. Klein and his group, and the support of the Alexander von Humboldt Foundation through a research fellowship. This work was partially supported by CONACyT and COSNET-SEP (Mexico).

REFERENCES

1. Pusey, P.N. and Tough, R.J.A., *Dynamic Light Scattering. Application of Photon Correlation Spectroscopy*, edited by R. Pecora (Plenum Press, 1985).
2. *Faraday Discussions of the Chemical Society* **76** (1983).
3. *Faraday Discussions of the Chemical Society* **83** (1987).
4. Medina-Noyola, M., *Faraday Discussions of the Chemical Society* **83**, 21 (1987).
5. Medina-Noyola, M. and Vizcarra-Rendon, A., *Phys. Rev.* **A32**, 3596 (1985); Ruiz-Estrada, H., Vizcarra-Rendon, A., Medina-Noyola, M. and Klein, R., *Phys. Rev.* **A34**, 3446 (1986); Vizcarra-Rendon, A., Ruiz-Estrada, H., Medina-Noyola, M. and Klein, R., *J. Chem. Phys.* **86**, 2976 (1987).
6. van Veluwen, A., Lekkerkerker, H.N.W., de Kruif, C.G. and Vrij, A., *Faraday Discussions of the Chemical Society* **83**, 59 (1987).
7. van Megen, W., Underwood, S.M., Ottewil, R.H., Williams, N.St.J. and Pusey, P.N., *Faraday Discussions of the Chemical Society* **83**, 47 (1987).
8. Batchelor, G.K., *J. Fluid Mech.* **74**, 1 (1976).
9. Felderhof, B.U., *Physica* **98A**, 373 (1977).
10. Beenakker, C.W.J. and Mazur, P., *Physica* **126A**, 349 (1984).
11. Medina-Noyola, M. and del Rio-Correa, J.L., *Physica* **146A**, 483 (1987).
12. Nägele, G., Medina-Noyola, M., Klein, R. and Arauz-Lara, J.L., *Physica* **149A**, 123 (1988).
13. Hess, W. and Klein, R., *Adv. Phys.* **32**, 173 (1983).
14. McQuarrie, D.A., *Statistical Mechanics* (Harper & Row, New York, 1976).
15. Hanna, S.N., Hess, W. and Klein, R., *Physica* **111A**, 181 (1982).
16. Ackerson, B.J. and Fleishman, L., *J. Chem. Phys.* **76**, 2675 (1982).
17. Lekkerkerker, H.N.W. and Dhont, J.K.G., *J. Chem. Phys.* **80**, 5790 (1984).
18. Medina-Noyola, M., *Phys. Rev. Letters* **60**, 2705 (1988).

LONGTIME, NONPREAVERAGED MOLECULAR DIFFUSIVITY, SEDIMENTATION VELOCITY AND TAYLOR DISPERSIVITY OF A FLUCTUATING CLUSTER OF INTERACTING BROWNIAN PARTICLES IN A VISCOUS FLOW

Howard Brenner
Department of Chemical Engineering,
Massachusetts Institute of Technology,
Cambridge, Massachusetts 02139, USA

Ali Nadim
Department of Mathematics,
Department of Chemical Engineering, Massachusetts Institute
of Technology, Cambridge, Massachusetts 02139, USA

and

Shimon Haber
Faculty of Mechanical Engineering,
Technion-Israel Institute of Technology, Haifa 32000, Israel

ABSTRACT. Generalized Taylor dispersion theory, incorporating so-called coupling effects, is used to calculate the transport properties of a single deformable 'chain' composed of hydrodynamically interacting rigid Brownian particles bound together by internal potentials, and moving through an unbounded quiescent viscous fluid. The individual rigid particles comprising the flexible chain or cluster may each be of arbitrary shape, size and density, and are assumed to be 'joined' together to form the chain by a configuration-dependent internal potential V. Each particle separately undergoes translational and rotational Brownian motions; together, their relative motions give rise to a conformational or vibrational Brownian motion of the chain (in additional to a translational motion of the chain as a whole). Sufficient time is allowed for all accessible chain configurations to be sampled many times in consequence of this internal Brownian motion. In consequence, an internal Boltzmann probabilistic distribution of conformations derived from V effectively obtains.

In contrast with prior analyses of such chain transport phenomena, no *ad hoc* pre-averaging hypothesis are invoked to effect the averaging of the input conformation-specific hydrodynamic mobility data. Rather, the calculation is effected rigorously within the usual (quasistatic) context of configuration-specific Stokes-Einstein equations.

Explicit numerical calculations serving to illustrate the general scheme are performed for the simplest case of dumbbells composed of identically-sized spheres connected by a slack tether.

1. INTRODUCTION

Transport mechanics[10, 11] in systems composed of isolated *rigid* particles moving through a continuous fluid phase has been the subject of extensive theoretical studies for well over a century. Building upon the pioneering hydrodynamic investigations of Stokes,[27] who examined the 'slow' viscous translational motion of a spherical particle through a quiescent Newtonian fluid, a field has emerged which includes a strikingly rich variety of phenomena. Classified under the general title of "low-Reynolds-number hydrodynamics",[16] the latter field incorporates such diverse areas as suspension rheology, sedimentation processes, translational and rotational Brownian motions, and colloid science —as well as a multitude of other nonequilibrium fluid-particle transport phenomena.

In circumstances where the suspended objects are *flexible* rather than rigid, progress has been limited, owing to the existence of several impediments. Not the least of these is the essentially pragmatic problem of dealing rigorously with the large numbers of degrees of freedom required to completely specify the instantaneous geometrical configuration of the flexible entity. A second related problem arises from the need to incorporate hydrodynamic interactions among the constituent rigid bodies making up the flexible body, and moving relative to one another. Usually, the first of these is dealt with within the more general framework of statistical mechanics[24] and kinetic theory,[5] while the second is circumvented by either the complete neglect of hydrodynamic interactions, or by invoking lower-order approximations, such as the Burgers-Oseen interaction tensors with preaveraging.[22] This apparent necessity for introducing approximate hydrodynamic interaction calculations into the requisite analysis has not only hindered quantitative progress in calculations pertaining to specific models, but also the actual conceptual development of existing theories. Thus, a major aim of the present study is to provide a fresh impetus to the rigorous theoretical development of macromolecular (flexible body) transport mechanics, by utilizing the newly developed framework of generalized Taylor dispersion theory[6, 8, 9] to complement classical kinetic treatments[3] of macromolecular hydrodynamics. Our proposed framework allows both of the aforementioned difficulties to be surmounted (at least conceptually); specifically, all translational and orientational degrees of freedom of the individual constituent rigid particles comprising the cluster are retained, as too are all the requisite, many-body, configuration-specific, hydrodynamic phenomenological coefficients (grand resistance and mobility matrices).

The flexible-body model addressed herein is assumed to consist of a *chain* or *cluster* of rigid particles, not unlike the classical 'bead-spring' models of Rouse[26] and Zimm.[31] However, in our treatment the constituent rigid particles are taken to be of finite size and to be of arbitrary shape; moreover, the configuration specific internal potential that serves to join them together —thereby permitting collective identification of the cluster or chain as a *single* entity moving through physical space— is assumed arbitrary (rather than being limited, for example, by such restrictions as pairwise additivity) so long as the potential is sufficiently attractive at large particle separations to assure convergence of any subsequent integrals that arise in the theory. Other common models, such as the 'bead-rod' models of Kramers[23] and Hassager,[18, 19] or those of 'segmentally flexible macromolecules',[17, 14, 29] can presumably be treated with appropriate choices of the potential, although quantum mechanical effects may unexpectedly arise[25] in effecting the transition from flexible to rigid form for the constraining potential.

In the realm of kinematics, any arbitrary motion of a rigid particle can be decomposed into a translation (of a locator point affixed to the particle) and a rigid-body rotation (about an axis through that point); however, the same is not true of the arbitrary motion of a flexible cluster of rigid bodies. Indeed, it is not *a priori* obvious which, if any, body-fixed geometrical point can best serve as a locator point for the chain 'position' in physical space. Points such as the centers of mass, volume, reaction,[4] diffusion[30] or even an arbitrary point affixed to any one of the constituent rigid particles, all appear to constitute equally reasonable candidates, although the ultimate physical results characterizing the long-time transport properties[13] of the cluster as a whole must necessarily show themselves to be independent of the explicit choice made for the body-fixed chain locator point.

Another element of interest, particularly in problems pertaining to the sedimentation of flexible chains,[32] is that although *on average* such a chain may possess a definite 'mean configuration', the chain may instantaneously exist in any one of an infinite number of other accessible geometrical configurations (with the probability of a specific configuration governed by a Boltzmann distribution, entailing the configuration-specific internal potential). For example, although on time average the flexible body may possess some definite symmetric shape, it does not generally possess this symmetry at all times or indeed at any single instant of time. Since such deviations from the 'average' configuration normally create long-time secular or cumulative effects, the long-time physical properties of such a body can be expected in general to differ from those of its symmetric, pre-averaged *rigid* counterpart. To rigorously analyze secular effects arising from instantaneous deviations from the average, generalized Taylor dispersion theory[6,13] can be employed. Indeed this paradigm has already been successfully used to investigate comparable sedimentation-dispersion phenomena in systems of *rigid* nonspherical particles[5,7] Upon incorporation of "coupling" effects[9] the generalized theory will be shown to be equally applicable to the macrotransport analysis of *flexible* clusters too.

Prediction of the conventional molecular diffusivity of flexible macromolecules[30,15] free of any sedimentation effects, is itself a challenging goal. Recently, Wegener[30] and Haber and Brenner[15] have independently recognized the important role of coupling among the translational, rotation and internal motions of flexible macromolecules. The former has shown via a perturbation analysis that the long-time macroscopic translational dispersivity of a flexible body is equal to the mean diffusivity of its (unique) center of diffusion. Haber and Brenner[15] had independently examined the same general problem within the Taylor-Aris dispersion[28,1,20] framework, performing detailed calculations for the case of a flexible dumbbell.

In the next section, a summary of the pertinent results of generalized coupled Taylor dispersion theory are furnished. Then, in section 3, transport mechanics of flexible chains and clusters —composed of rigid Brownian particles and sedimenting through an otherwise quiescent viscous fluid— are reviewed. The results obtained through application of generalized coupled Taylor dispersion theory are outlined, and specific numerical calculations pertaining to simple tethered dumbbells presented. Our general aim in the presentation of this work is to highlight and outline the principal methodology of our scheme. The details, which have deliberately been omitted, may be found in a concurrent contribution,[12] of which the present paper constitutes a mere gloss.

2. GENERALIZED COUPLED TAYLOR DISPERSION THEORY

A significant portion of the effort required to understand the non-equilibrium behavior of suspensions of flexible chains must necessarily be devoted to first describing the transport of a single one of the flexible bodies within the phase-space codifying its internal and external configurations. In addition to the conventional convective and (direct) diffusive contributions, the transport of isolated flexible objects also entails indirect or coupling contributions, whereby probability density gradients in the internal space give rise to fluxes in the external space and conversely. In order to treat the effects of such coupling upon the macrotransport properties of the isolated body (and ultimately upon the macroscale properties of the suspension as a whole), the rigorous results of generalized Taylor dispersion theory[6,8] —including coupling effects[9]— will be employed. Accordingly, this section is devoted to reviewing the relevant results of the latter theory. Subsequently, in the next section, the theory will be applied to the important problem of Brownian diffusion and sedimentation of a flexible chain or cluster, composed of hydrodynamically-interacting rigid Brownian particles, and moving through a quiescent viscous fluid.

The starting point of generalized Taylor dispersion theory is the canonical continuity equation[9]

$$\frac{\partial P}{\partial t} + \nabla_Q \cdot \mathbf{J} + \nabla_q \cdot \mathbf{j} = \delta(t)\delta(\mathbf{Q}-\mathbf{Q}')\delta(\mathbf{q}-\mathbf{q}') \quad (1)$$

governing conservation and transport of the conditional probability density $P(\mathbf{Q},\mathbf{q},t\mid \mathbf{Q}',\mathbf{q}')$ for a flexible 'tracer' to possess internal coordinates $\mathbf{q}$ and external (physical-space) coordinates **Q** at the instant t, given that at time $t = 0$ its respective $(\mathbf{Q}, \mathbf{q})$ coordinates were $(\mathbf{Q}', \mathbf{q}')$. The local (internal) and global (external) fluxes **j** and **J** are respectively assumed to be given by the constitutive expressions

$$\mathbf{j} = \mathbf{u}(\mathbf{q})P - \mathbf{D}^{qQ}(\mathbf{q}) \cdot \nabla_Q P - e^{-E}\mathbf{D}^{qq}(\mathbf{q}) \cdot \nabla_q(e^E P), \tag{2}$$

$$\mathbf{J} = \mathbf{U}(\mathbf{q})P - \mathbf{D}^{QQ}(\mathbf{q}) \cdot \nabla_Q P - e^{-E}\mathbf{D}^{Qq}(\mathbf{q}) \cdot \nabla_q(e^E P). \tag{3}$$

In addition to the usual *direct* contributions arising from convection and diffusion in each subspace (*i.e.*, local and global), coupling effects between these subspaces have been included, as directly embodied in the existence of the coupling diffusivity $\mathbf{D}^{qQ}$ (or its transpose $\mathbf{D}^{Qq}$). The Fickian diffusivities $\mathbf{D}^{QQ}$ and $\mathbf{D}^{qq}$ are each assumed to be both symmetric and positive definite, as too is also the combination $\mathbf{D}^{QQ} - \mathbf{D}^{Qq} \cdot (\mathbf{D}^{qq})^{-1} \cdot \mathbf{D}^{qQ}$. Inclusion of the dimensionless local-space potential $E(\mathbf{q})$ in the form given above summarizes the transport effects due to conservative forces;[6, 8] **U** and **u** are the respective global —and local— space velocity vectors of the tracer.

The conservation equation (1) is to be solved subject to appropriate attenuation-rate boundary conditions imposed as $|Q| \to \infty$ to assure convergence, a no-flux condition $\mathbf{n} \cdot \mathbf{j} = 0$ at the boundaries ∂q_0 of the local space, and the (pre-) initial condition $P = 0$ for $t < 0$. Application of a Lagrangian moment analysis to these equations reveals that the purely global transport of the tracer (*i.e.*, from which transport process the internal degrees of freedom **q** have been eliminated via integration) is characterized by two macrotransport coefficients, namely the mean tracer velocity vector through physical space,

$$\overline{\mathbf{U}}^* = \int_{q_0} d\mathbf{q}\, [P_0^\infty \mathbf{U} - e^{-E}\mathbf{D}^{Qq} \cdot \nabla_q(e^E P_0^\infty)], \tag{4}$$

and its physical-space dispersivity dyadic,

$$\overline{\mathbf{D}}^* = \int_{q_0} d\mathbf{q}\, \left\{ P_0^\infty [\mathbf{D}^{QQ} - \mathbf{D}^{Qq} \cdot \nabla_q \mathbf{B}] + [P_0^\infty(\mathbf{U} - \overline{\mathbf{U}}^*) - e^{-E}\mathbf{D}^{Qq} \cdot \nabla_q(e^E P_0^\infty)]\mathbf{B} \right\}. \tag{5}$$

The pair of coefficient $\overline{\mathbf{U}}^*$ and $\overline{\mathbf{D}}^*$ serves to describe the **Q**-space kinematical properties of the flexible cluster for times sufficiently long to have established a terminal state (equilibrium or quasi-steady) within the internal space, a state that is independent of the initial internal configuration $\mathbf{q}'$.

Calculation of these long-time macrotransport coefficients requires knowledge of the internal equilibrium density P_0^∞, satisfying[9]

$$\nabla_q \cdot [\mathbf{u}P_0^\infty - e^{-E}\mathbf{D}^{qq} \cdot \nabla_q(e^E P_0^\infty)] = 0, \tag{6a}$$

$$\mathbf{n} \cdot [\mathbf{u}P_0^\infty - e^{-E}\mathbf{D}^{qq} \cdot \nabla_q(e^E P_0^\infty)] = 0 \quad \text{on} \quad \partial q_0 \tag{6b}$$

$$\int_{q_0} d\mathbf{q}\, P_0^\infty = 1. \tag{6c}$$

Also required in (5) is the solution $\mathbf{B}(\mathbf{q})$ of the equations

$$\nabla_q \cdot [\mathbf{u}P_0^\infty \mathbf{B} - e^{-E}\mathbf{D}^{qq} \cdot \nabla_q(e^E P_0^\infty \mathbf{B}) + P_0^\infty \mathbf{D}^{qQ}] \tag{7a}$$
$$= P_0^\infty(\mathbf{U} - \overline{\mathbf{U}}^*) - e^{-E}\mathbf{D}^{Qq} \cdot \nabla_q(e^E P_0^\infty),$$

$$P_0^\infty \mathbf{n} \cdot [\mathbf{D}^{qq} \cdot \nabla_q \mathbf{B} - \mathbf{D}^{qQ}] = 0 \quad \text{on} \quad \partial q_0, \tag{7b}$$

uniquely defining **B** to within an irrelevant arbitrary additive constant vector.[9]

Since only the long-time **Q**-space transport of the tracer is generally of interest in practice, the simplification afforded by being able to characterize the macrotransport process via only the two position-independent phenomenological coefficients (4) and (5) —in place of the much larger set of **q**-dependent phenomenological coefficients appearing in (2)–(3)— represents a major achievement of the general theory [*albeit* the not inconsiderable effort perhaps necessary to actually solve (6) and (7)].

3. BROWNIAN DIFFUSION AND SEDIMENTATION OF FLEXIBLE CHAINS AND CLUSTERS

Consider a flexible cluster, synthesized by joining together $n+1$ rigid particles of arbitrary shapes via interparticle (internal) potentials. These potentials, which can be as elementary as simple tethers connecting pairs of particles, or as complex as one may wish to imagine, serve to permit collective identification of the $n+1$ rigid particles as a *single* identifiable entity —namely, a "flexible chain". Its 'flexibility' arises from the fact that its constituent rigid particles are free to translate and rotate relative to one another, subject to any configurational constraints imposed by the internal potential; as such, the conformation of the chain can (and does) vary with time.

Label the individual rigid particles with the index A ($A = 0, 1, 2, \ldots, n$, for a total of $n+1$ particles), and denote by O_A an arbitrarily-positioned particle 'locator point', rigidly affixed to particle A. At each instant the complete configuration of the cluster is entirely determined by specification of the $n+1$ position vectors $\mathbf{R}_A$ of the locator points O_A relative to an arbitrary space-fixed origin, and a comparable set of $n+1$ orientational triplets $\boldsymbol{\phi}_A$ of each of the constituent particles.

Denote by

$$P(\mathbf{R}_0, \ldots, \mathbf{R}_n; \boldsymbol{\phi}_0, \ldots, \boldsymbol{\phi}_n, t \mid \mathbf{R}'_0, \ldots, \mathbf{R}'_n; \boldsymbol{\phi}'_0, \ldots, \boldsymbol{\phi}'_n) \tag{8}$$

the conditional probability density for finding the chain configuration in an infinitesimal neighborhood of $(\mathbf{R}_0, \ldots, \mathbf{R}_n; \boldsymbol{\phi}_0, \ldots, \boldsymbol{\phi}_n)$ at time t, given that at time $t=0$ the chain possessed the corresponding primed configuration. The probability density P is chosen to satisfy the dual requirements

$$\int d\mathbf{R}_0 \cdots d\mathbf{R}_n d\boldsymbol{\phi}_0 \cdots d\boldsymbol{\phi}_n\, P = 1 \qquad (t > 0), \tag{9a}$$

$$P = 0 \qquad (t < 0), \tag{9b}$$

in which the limits of integration extend over the entire physical and orientational-space domains available for the configurational transport.

The conservation equation governing the detailed configurational transport of the flexible cluster through a fluid continuum is of the general form

$$\frac{\partial P}{\partial t} + \sum_{A=0}^{n} \left(\frac{\partial}{\partial \mathbf{R}_A} \cdot \mathbf{J}[\mathbf{R}_A] + \frac{\partial}{\partial \boldsymbol{\phi}_A} \cdot \mathbf{j}[\boldsymbol{\phi}_A] \right) = \delta(t) \prod_{A=0}^{n} \delta(\mathbf{R}_A - \mathbf{R}'_A) \delta(\boldsymbol{\phi}_A - \boldsymbol{\phi}'_A). \tag{10}$$

The configuration-specific physical —and orientational— space vector flux densities $\mathbf{J}[\mathbf{R}_A]$ and $\mathbf{j}[\boldsymbol{\phi}_A]$ of particle A will be assumed to possess the respective convective-diffusive constitutive forms

$$\mathbf{J}[\mathbf{R}_A] = \dot{\mathbf{R}}_A P - \sum_{B=0}^{n} \left\{ \mathbf{D}[\mathbf{R}_A \mid \mathbf{R}_B] \cdot \frac{\partial P}{\partial \mathbf{R}_B} + \mathbf{D}[\mathbf{R}_A \mid \boldsymbol{\phi}_B] \cdot \frac{\partial P}{\partial \boldsymbol{\phi}_B} \right\}, \tag{11a}$$

$$\mathbf{j}[\boldsymbol{\phi}_A] = \dot{\boldsymbol{\phi}}_A P - \sum_{B=0}^{n} \left\{ \mathbf{D}[\boldsymbol{\phi}_A \mid \mathbf{R}_B] \cdot \frac{\partial P}{\partial \mathbf{R}_B} + \mathbf{D}[\boldsymbol{\phi}_A \mid \boldsymbol{\phi}_B] \cdot \frac{\partial P}{\partial \boldsymbol{\phi}_B} \right\}, \tag{11b}$$

in which

$$\dot{\mathbf{R}}_A = -\sum_{B=0}^{n} \left\{ \mathbf{M}[\mathbf{R}_A \mid \mathbf{R}_B] \cdot \frac{\partial V}{\partial \mathbf{R}_B} + \mathbf{M}[\mathbf{R}_A \mid \boldsymbol{\phi}_B] \cdot \frac{\partial V}{\partial \boldsymbol{\phi}_B} \right\}, \tag{12a}$$

$$\dot{\boldsymbol{\phi}}_A = -\sum_{B=0}^{n} \left\{ \mathbf{M}[\boldsymbol{\phi}_A \mid \mathbf{R}_B] \cdot \frac{\partial V}{\partial \mathbf{R}_B} + \mathbf{M}[\boldsymbol{\phi}_A \mid \boldsymbol{\phi}_B] \cdot \frac{\partial V}{\partial \boldsymbol{\phi}_B} \right\}. \tag{12b}$$

The configuration-specific mobility dyadics $\mathbf{M}[A \mid B]$ and diffusivity dyadics $\mathbf{D}[A \mid B]$ are functionally dependent only upon the internal configuration of the cluster; they are closely related to one another via the multibody configuration-specific Stokes-Einstein relations[4]

$$\mathbf{D}[A \mid B] = kT\mathbf{M}[A \mid B]. \tag{13}$$

The potential $V(\mathbf{R}_0, \ldots, \mathbf{R}_n; \boldsymbol{\phi}_0, \ldots, \boldsymbol{\phi}_n)$ appearing in (12) is also assumed to be a known function of the internal configuration, to which potential is added a further contribution arising from a constant external force $\mathbf{F}$ (e.g., gravity) causing sedimentation of the chain [cf. (16)] as a whole.

Subject to appropriate boundary and initial conditions, equation (10) for P may be solved to obtain an exact description of the configurational transport process; however, such a detailed resolution of the problem is not ordinarily the ultimate objective of interest. Indeed, for large n, such a description would be overwhelmingly detailed. Rather, if the flexible object is to be viewed as an entity unto itself, a much more physically useful and concise description is that of the transport through physical space of the flexible body as a whole, viewed as the sole object of interest. Attainment of this goal requires that we assign a single locator point to the flexible cluster as a definable entity, and focus exclusively on the stochastic trajectory of that point through ordinary three-dimensional (i.e., physical) space. All other coordinates, aside from the three required to specify the position of that locator point in physical space, are then assigned the role of specifying the internal conformation of the chain. Our eventual goal is thus to eliminate from the transport equation all the internal degrees of freedom, at least for sufficiently long times to assure that equilibrium with respect to the internal conformation has effectively been attained.

Equations (10)–(12), by themselves, effect no definite decomposition of the internal variables into respective local (internal) and global (external) sets. As this classification is prerequisite in applying to particular problems the results of coupled generalized Taylor dispersion theory given in section 2, we will presently effect an appropriate transformation permitting this classification to be made.

Local/global form of the equations

Our *ansatz* consists of recognizing that *for long times* the choice of locator point for the flexible cluster is irrelevant; that is, any and all choices of chain locator point suffice equally well for identifying the location of the flexible body in physical space. In the long run, all such points behave alike as regards their *net* translational motions in physical space. As such, we are free to choose any point O_0 rigidly affixed to particle $A = 0$ as the locator point of the flexible cluster. Denote by $\mathbf{Q}$ the position vector of this point, so that

$$\mathbf{Q} \equiv \mathbf{R}_0, \tag{14}$$

spanning the entire global (physical) space available for chain transport. Consider the local variables q to consist of the set

$$\mathbf{q} \equiv (\mathbf{r}_1, \mathbf{r}_2, \ldots, \mathbf{r}_n, \boldsymbol{\phi}_0, \boldsymbol{\phi}_1, \ldots, \boldsymbol{\phi}_n) \equiv (\mathbf{r}^n, \boldsymbol{\phi}^{n+1}), \tag{15a}$$

in which

$$\mathbf{r}_a \equiv \mathbf{R}_a - \mathbf{R}_0 \quad (a = 1, 2, \ldots, n) \tag{15b}$$

are the position vectors of the remaining points O_a relative to the designated point O_0.

Effecting the transformation of coordinates from $(\mathbf{R}_0, \ldots, \mathbf{R}_n; \boldsymbol{\phi}_0, \ldots, \boldsymbol{\phi}_n)$ to $(\mathbf{Q}, \mathbf{q})$ —as represented by (14) and (15)— into (10)–(12), in conjunction with the assumed decomposition of V into the global $\oplus$ local form

$$V = -\mathbf{F} \cdot \mathbf{R}_0 + kTE(\mathbf{r}^n, \boldsymbol{\phi}^{n+1}), \tag{16}$$

(with $\mathbf{F}$ a constant external force vector), ultimately yields the global/local form

$$\frac{\partial P}{\partial t} + \boldsymbol{\nabla}_Q \cdot \mathbf{J} + [\![\boldsymbol{\nabla}_q]\!]^\dagger \cdot [\![\mathbf{j}]\!] = \delta(t)\delta(\mathbf{Q} - \mathbf{Q}')\delta(\mathbf{q} - \mathbf{q}'), \tag{17}$$

of the conservation equation for P, in which

$$[\![\boldsymbol{\nabla}_q]\!]^\dagger \equiv [\partial/\partial \mathbf{r}_1^\dagger \cdots \partial/\partial \mathbf{r}_n^\dagger \; \partial/\partial \boldsymbol{\phi}_0^\dagger \cdots \partial/\partial \boldsymbol{\phi}_n^\dagger] \tag{18}$$

$$[\![\mathbf{j}]\!] = [\![\mathbf{u}]\!]P - [\![\mathbf{D}^{qQ}]\!] \cdot \boldsymbol{\nabla}_Q P - e^{-E}[\![\mathbf{D}^{qq}]\!] \cdot [\![\boldsymbol{\nabla}_q]\!](e^E P), \tag{19}$$

$$\mathbf{J} = \mathbf{U}P - \mathbf{D}^{QQ} \cdot \boldsymbol{\nabla}_Q P - e^{-E}[\![\mathbf{D}^{Qq}]\!] \cdot [\![\boldsymbol{\nabla}_q]\!](e^E P). \tag{20}$$

In the above, the double-bracketed quantities are partitioned matrices, whose individual elements are themselves vectors or dyadics. the phenomenological coefficients appearing in (19) and (20) are defined as follows:

(i) *Local velocity matrix* $[\![\mathbf{u}]\!]$:

$$[\![\mathbf{u}]\!] = \left[\!\!\left[\begin{array}{c} \mathbf{u}[\mathbf{r}_a] \\ \mathbf{u}[\boldsymbol{\phi}_A] \end{array} \right]\!\!\right], \tag{21a}$$

with

$$\mathbf{u}[\mathbf{r}_a] = \{\mathbf{M}[\mathbf{R}_a \mid \mathbf{R}_0] - \mathbf{M}[\mathbf{R}_0 \mid \mathbf{R}_0]\} \cdot \mathbf{F}, \tag{21b}$$

$$\mathbf{u}[\boldsymbol{\phi}_A] = \mathbf{M}[\boldsymbol{\phi}_A \mid \mathbf{R}_0] \cdot \mathbf{F}; \tag{21c}$$

(ii) *Coupling diffusivity matrices* $[\mathbf{D}^{qQ}]$ and $[\mathbf{D}^{Qq}]$:

$$[\mathbf{D}^{qQ}] = [\mathbf{D}^{Qq}]^\dagger = \left[\begin{bmatrix} \mathbf{D}^{qQ}[\mathbf{r}_a] \\ \mathbf{D}^{qQ}[\boldsymbol{\phi}_A] \end{bmatrix}\right], \tag{22a}$$

with

$$\mathbf{D}^{qQ}[\mathbf{r}_a] = \mathbf{D}[\mathbf{R}_a \mid \mathbf{R}_0] - \mathbf{D}[\mathbf{R}_0 \mid \mathbf{R}_0], \tag{22b}$$

$$\mathbf{D}^{qQ}[\boldsymbol{\phi}_A] = \mathbf{D}[\boldsymbol{\phi}_A \mid \mathbf{R}_0]; \tag{22c}$$

(iii) *Local diffusivity matrix* $[\mathbf{D}^{qq}]$:

$$[\mathbf{D}^{qq}] = \left[\begin{bmatrix} \mathbf{D}^{qq}[\mathbf{r}_a \mid \mathbf{r}_b] & \mathbf{D}^{qq}[\mathbf{r}_a \mid \boldsymbol{\phi}_A] \\ \mathbf{D}^{qq}[\boldsymbol{\phi}_A \mid \mathbf{r}_a] & \mathbf{D}^{qq}[\boldsymbol{\phi}_A \mid \boldsymbol{\phi}_B] \end{bmatrix}\right], \tag{23a}$$

in which

$$\mathbf{D}^{qq}[\mathbf{r}_a \mid \mathbf{r}_b] = \mathbf{D}[\mathbf{R}_a \mid \mathbf{R}_b] - \mathbf{D}[\mathbf{R}_a \mid \mathbf{R}_0] - \mathbf{D}[\mathbf{R}_0 \mid \mathbf{R}_b] + \mathbf{D}[\mathbf{R}_0 \mid \mathbf{R}_0], \tag{23b}$$

$$\mathbf{D}^{qq}[\mathbf{r}_a \mid \boldsymbol{\phi}_A] = \mathbf{D}[\mathbf{R}_a \mid \boldsymbol{\phi}_A] - \mathbf{D}[\mathbf{R}_0 \mid \boldsymbol{\phi}_A], \tag{23c}$$

$$\mathbf{D}^{qq}[\boldsymbol{\phi}_A \mid \mathbf{r}_a] = \mathbf{D}[\boldsymbol{\phi}_A \mid \mathbf{R}_a] - \mathbf{D}[\boldsymbol{\phi}_A \mid \mathbf{R}_0] = \mathbf{D}^{qq\dagger}[\mathbf{r}_a \mid \boldsymbol{\phi}_A], \tag{23d}$$

$$\mathbf{D}^{qq}[\boldsymbol{\phi}_A \mid \boldsymbol{\phi}_B] = \mathbf{D}[\boldsymbol{\phi}_A \mid \boldsymbol{\phi}_B]; \tag{23e}$$

(iv) *Global velocity vector* $\mathbf{U}$:

$$\mathbf{U} \equiv \mathbf{M}[\mathbf{R}_0 \mid \mathbf{R}_0] \cdot \mathbf{F}; \tag{24}$$

(v) *Global diffusivity dyadic* $\mathbf{D}^{QQ}$:

$$\mathbf{D}^{QQ} \equiv \mathbf{D}[\mathbf{R}_0 \mid \mathbf{R}_0]. \tag{25}$$

Equations (17), (19) and (20) are identical to the canonical forms of the equations of generalized coupled Taylor dispersion theory (1)–(3), albeit the more complex appearance of their partitioned-matrix phenomenological coefficients. As such, the results (4) and (5) are directly applicable in this circumstance.

4. RESULTS

For completeness, explicit expressions for the mean velocity vector $\overline{\mathbf{U}}^*$ and dispersivity dyadic $\overline{\mathbf{D}}^*$ of the flexible chain sedimenting through an otherwise quiescent viscous fluid are now presented. These correspond identically to equations (4) and (5) of section 2, namely

$$\overline{\mathbf{U}}^* = \int_{q_0} d\mathbf{q}\, \{P_0^\infty \mathbf{U} - e^{-E}[\mathbf{D}^{Qq}] \cdot [\boldsymbol{\nabla}_q](e^E P_0^\infty)\} \tag{26}$$

and

$$\overline{\mathbf{D}}^* = \int_{q_0} d\mathbf{q} \left\{ P_0^\infty [\mathbf{D}^{QQ} - [\mathbf{D}^{Qq}] \cdot [\boldsymbol{\nabla}_q] \mathbf{B}] \right. \tag{27}$$
$$\left. + [P_0^\infty (\mathbf{U} - \overline{\mathbf{U}}^*) - e^{-E}[\mathbf{D}^{Qq}] \cdot [\boldsymbol{\nabla}_q](e^E P_0^\infty)] \mathbf{B} \right\}.$$

The requisite local fields P_0^∞ and $\mathbf{B}$ appearing above are to be obtained by solving equations equivalent to (6) and (7).

Expressions (26) and (27) represent the major results of our analysis. Together with identification from existing data of the phenomenological coefficients required therein, as given by (21)–(25), they permit rigorous calculation of the long-time mean sedimentation velocity vector $\overline{\mathbf{U}}^*$ and dispersivity dyadic $\overline{\mathbf{D}}^*$ for any flexible chain or cluster composed of rigid Brownian constituent particles bound together via internal potentials.

5. EXAMPLE

As an example consider the sedimentation-dispersion problem for the tethered dumbbell depicted in Figure 1. The latter consists of two identical rigid spherical particles of radii a, each denser than its surrounding fluid by an amount corresponding to a mass difference $|\Delta m|$. A tether of ultimate extensibility L is assumed to connect their geometric centers. The total sedimentation force acting on the dumbbell is here

$$\mathbf{F} = 2|\Delta m|\mathbf{g} = F\hat{\mathbf{F}}, \tag{28}$$

with $\hat{\mathbf{F}}$ a unit vector along gravity field vector $\mathbf{g}$. If the dimensionless tether-length/sphere-radius ratio is denoted by

$$\chi = L/a, \tag{29}$$

and the Langevin parameter (embodying the ratio of gravitational to thermal energies of the dumbbell) designated as

$$\gamma = Fa/kT, \tag{30}$$

the final results of the foregoing analysis may be summarized by the explicit relations

$$\frac{\overline{\mathbf{U}}^*}{M_\infty F} = \widehat{\overline{M}} \hat{\mathbf{F}} \tag{31}$$

and

$$\frac{\overline{\mathbf{D}}^*}{D_\infty} = \hat{D}_\| \hat{\mathbf{F}}\hat{\mathbf{F}} + \hat{D}_\perp (\mathbf{I} - \hat{\mathbf{F}}\hat{\mathbf{F}}), \tag{32a}$$

in which

$$\hat{D}_\| = \widehat{\overline{M}} + \gamma^2 (\hat{\alpha} + \hat{\beta}/5), \tag{32b}$$
$$\hat{D}_\perp = \widehat{\overline{M}} + \gamma^2 (3\hat{\beta}/20). \tag{32c}$$

Here,

$$M_\infty = (12\pi\mu a)^{-1}, \qquad D_\infty = kT M_\infty \tag{33a,b}$$

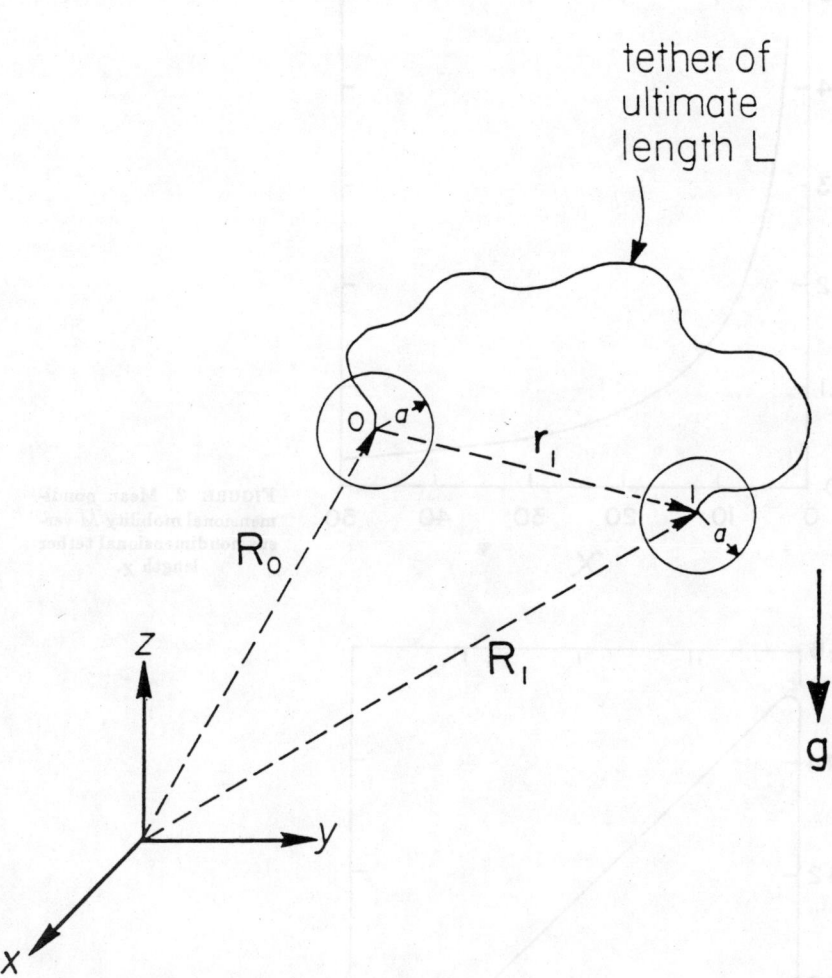

FIGURE 1. Tethered dumbbell sedimenting under the influence of gravity (g) in an otherwise quiescent viscous fluid.

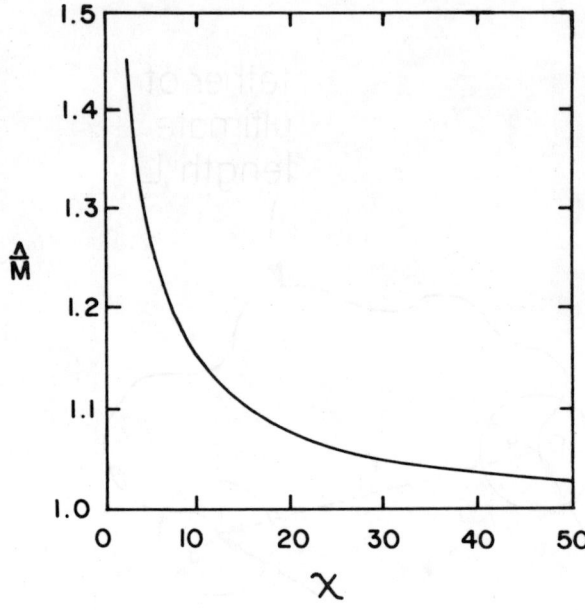

FIGURE 2. Mean nondimensional mobility $\widehat{\overline{M}}$ versus nondimensional tether length χ.

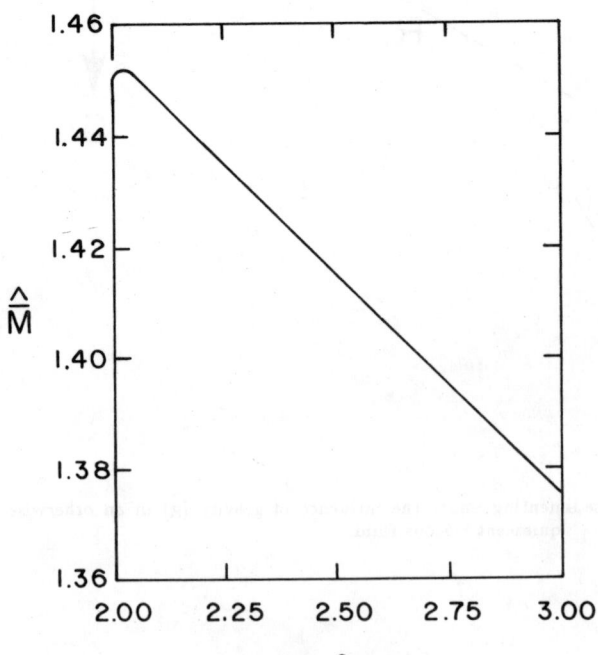

FIGURE 3. Mean nondimensional mobility $\widehat{\overline{M}}$ versus nondimensional tether length χ in the near touching range $2 \leq \chi \leq 3$.

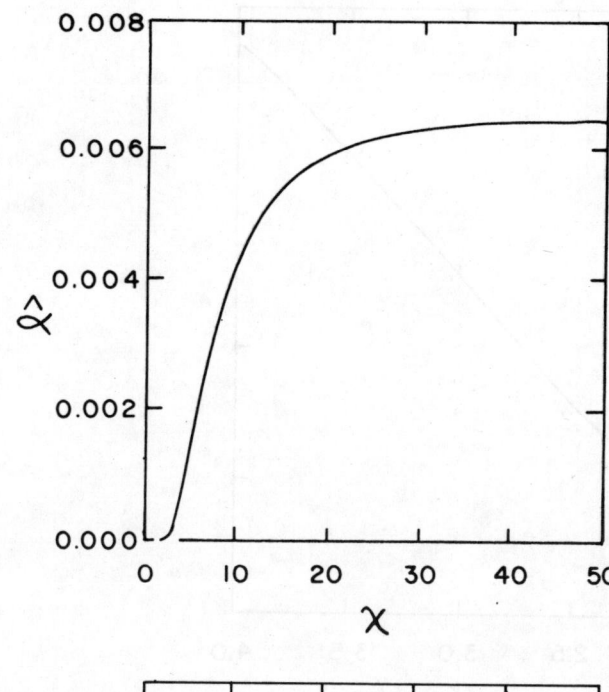

FIGURE 4. Taylor dispersivity coefficient $\hat{\alpha}$ versus nondimensional tether length χ.

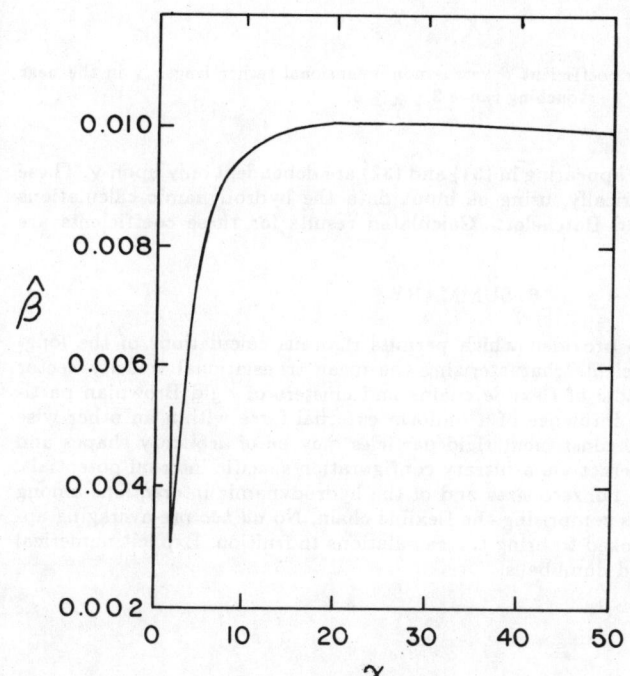

FIGURE 5. Taylor dispersivity coefficient $\hat{\beta}$ versus nondimensional tether length χ.

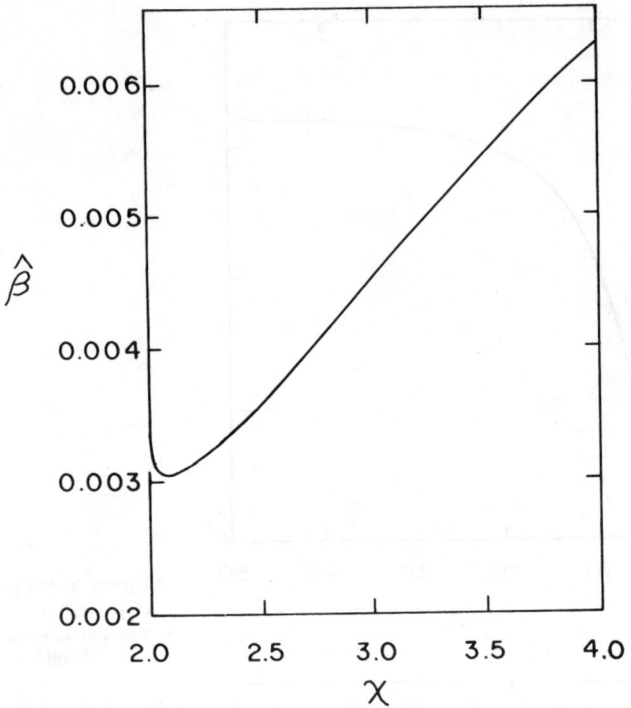

FIGURE 6. Taylor dispersivity coefficient $\hat{\beta}$ versus nondimensional tether length χ in the near touching range $2 \leq \chi \leq 4$.

with μ the fluid viscosity.

The functions $\widehat{\overline{M}}$, $\hat{\alpha}$ and $\hat{\beta}$ appearing in (31) and (32) are dependent only upon χ. These have been calculated numerically, using as input data the hydrodynamic calculations of Jeffrey and Onishi[21] and Batchelor[2] Calculated results for these coefficients are embodied in Figures 2–6.

6. SUMMARY

A general scheme has been provided which permits rigorous calculations of the long-time macrotransport coefficients characterizing the mean translational velocity vector and Taylor dispersivity dyadic of flexible chains and clusters of rigid Brownian particles, sedimenting under the influence of a uniform external force within an other wise quiescent viscous fluid. The constituent rigid particles may be of arbitrary shapes and sizes, and may mutually interact via arbitrary configuration-specific internal potentials. Full account is taken of the nonzero sizes and of the hydrodynamic interactions among the individual rigid particles comprising the flexible chain. No *ad hoc* pre-averaging approximations have been invoked to bring the calculations to fruition. Explicit numerical results are given for tethered dumbbells.

ACKNOWLEDGEMENTS

H.B. would like to thank the Organizing Committee of the XVII Winter Meeting on Statistical Physics, including Drs. Enrique Fernandez, Carmen Varea, Agustin Gonzalez and Esteban Martina for their kind invitation to participate and hospitality at Oaxtepec. This work was supported by grants from the Office of Basic Energy Sciences of the Department of Energy and the National Science Foundation.

REFERENCES

1. Aris, R., "On the dispersion of a solute in a fluid flowing through a tube", *Proc. Roy. Soc.* **A235**, 67–77 (1956).
2. Batchelor, G.K., "Brownian diffusion of particles with hydrodynamic interaction", *J. Fluid Mech.* **74**, 1–29 (1976).
3. Bird, R.B., Hassager, O., Armstrong, R.C. and Curtiss, C.F., *Dynamics of Polymeric Liquids, Vol. 2, Kinetic Theory* (Wiley & Sons, 1977).
4. Brenner, H., "Coupling between the translational and rotational Brownian motions of rigid particles of arbitrary shape II. General theory", *J. Colloid Interface Sci.* **23**, 407–436 (1967).
5. Brenner, H., "Taylor dispersion in systems of sedimenting non-spherical Brownian particles I. Homogeneous, centrosymmetric, axisymmetric particles", *J. Colloid Interface Sci.* **71**, 189–208 (1979).
6. Brenner, H., "A general theory of Taylor dispersion phenomena", *PhysicoChem. Hydrodyn.* **1**, 19–23 (1980).
7. Brenner, H., "Taylor dispersion in systems of sedimenting non-spherical Brownian particles II. Homogeneous ellipsoidal particles", *J. Colloid Interface Sci.* **80**, 548–588 (1981).
8. Brenner, H., "A general theory of Taylor dispersion phenomena II. An extension", *PhysicoChem. Hydrodyn.* **3**, 139–157 (1982).
9. Brenner, H., "A general theory of Taylor dispersion phenomena IV. Direct coupling effects", *Chem. Eng. Comm.* **18**, 355–379 (1982).
10. Brenner, H. and Condiff, D.W., "Transport mechanics in systems of orientable particles III. Arbitrary particles", *J. Colloid Interface Sci.* **41**, 228–274 (1972).
11. Brenner, H. and Condiff, D.W., "Transport mechanics in systems of orientable particles IV. Convective transport", *J. Colloid Interface Sci.* **47**, 199–264 (1974).
12. Brenner, H., Nadim, A. and Haber, S., "Long-time molecular diffusion, sedimentation and Taylor dispersion of a fluctuating cluster of interacting particles", *J. Fluid Mech.* **183**, 511–542 (1987).
13. Brenner, H. and Pagitsas, M., *Macrotransport Processes* (to appear, 1987).
14. Garcia de la Torre, J., Mellado, P. and Rodes, V., "Diffusion coefficients of segmentally flexible macromolecules with two spherical subunits", *Biopolymers* **24**, 2145–2164 (1985).
15. Haber, S. and Brenner, H., "Taylor-Aris dispersion of N particles with internal constraints I. Molecular dispersion, *recent Developments in Structured Continua*, P.N. Kaloni and D. DeKee, Eds., Longmans, 1986.
16. Happel, J. and Brenner, H., *Low Reynolds Number Hydrodynamics*, Nijhoff, 1983.
17. Harvey, S.C., Mellado, P. and Garcia de la Torre, J., "Hydrodynamic resistance and diffusion coefficients of segmentally flexible macromolecules with two subunits", *J. Chem. Phys.* **78**, 2081–2090 (1983).
18. Hassager, O., "Kinetic theory and theology of bead-rod models for macromolecular solutions I. Equilibria and steady flow properties", *J. Chem. Phys.* **60**, 2111–2124 (1974).

19. Hassager, O., "Kinetic theory and rheology of bead-rod models for macromolecular solutions II. Linear unsteady flow properties"', *J. Chem. Phys.* **60**, 4001–4008 (1974).
20. Horn, F.J.M., "Calculation of dispersion coefficients by means of moments", *AIChE J.* **17**, 613–620 (1971).
21. Jeffrey, D.J. and Onishi, Y., "Calculation of the resistance and mobility functions for two unequal spheres in low-Reynolds-number flow", *J. Fluid Mech.* **139**, 261–290 (1984).
22. Kirkwood, J.G. and Riseman, J., "The intrinsic viscosities and diffusion constants of flexible macromolecules in solution", *J. Chem. Phys.* **16**, 565–573 (1948).
23. Kramers, H.A., "The behavior of macromolecules in inhomogeneous flow", *J. Chem. Phys.* **14**, 415–424 (1946).
24. Landau, L.D. and Lifshitz, E.M., *Course of Theoretical Physics Vol. 5, Statistical Mechanics* (Pergamon, 1980).
25. Rallison, J.M., "The role of rigidity constraints in the rheology of dilute polymer solutions", *J. Fluid Mech.* **93**, 251–279 (1979).
26. Rouse, P.E., "A theory of linear viscoelastic properties of coiling polymers", *J. Chem. Phys.* **21**, 1272–1280 (1953).
27. Stokes, G.G., "On the effect of the internal friction of fluids on the motion of pendulums", *Trans. Cambr. Phil. Soc.* **9** 8–106 (1951).
28. Taylor, G.I., "Dispersion of soluble matter in solvent flowing slowly through a tube", *Proc. Roy. Soc.* **A219**, 186–203 (1953).
29. Wegener, W.A., "Bead models of segmentally flexible macromolecules", *J. Chem. Phys.* **76**, 6425–6430 (1982).
30. Wegener, W.A., "Center of diffusion of flexible macromolecules", *Macromolecules* **18**, 2522–2530 (1985).
31. Zimm, B.H., "Dynamics of polymer molecules in dilute solution: visco-elasticity, flow birefringence and dielectric loss", *J. Chem. Phys.* **24**, 269–278 (1956).
32. Zimm, B.H., "Sedimentation of asymmetric elastic dumbbells and the rigid-body approximation in the hydrodynamics of chains", *Macromolecules* **15**, 520–525 (1982).

SIMPLE MODELS FOR DIFFUSION LIMITED REACTIONS

F. Leyvraz

Laboratorio de Cuernavaca, Instituto de Física,
Universidad Nacional Autónoma de México,
apartado postal 20-364, 01000 México, D.F. MEXICO

ABSTRACT. We discuss the asymptotic long-time behavior of various systems characterized by their strongly non-equilibrium nature, leading to the absence of detailed balance or sometimes even of a non-trivial equilibrium. It is shown that, when the rate-limiting step is assumed to be the transport mechanism (rather than the reaction rate), anomalous behavior usually occurs in low dimensions, due to the build-up of ever longer range spatial correlations. We shall only investigate here the case where the transport mechanism is ordinary diffusion. We further show that for higher dimensions, the classical picture provided by the rate equation approach is qualitatively correct.

1. INTRODUCTION

In many physical systems, various types of irreversible processes take place that are of paramount importance.In the following, we shall focus on a rather specific and widespread type of irreversible process that we shall call reactions, even though most of the examples are not drawn from chemistry. As a first instance, let us consider aggregation of colloids[1] in the limit where the cluster size is not sufficiently large to influence the rate at which clusters react. In this case, the process could be summarized by the following reaction scheme:

$$A + A \xrightarrow{k} A \qquad (1)$$

where A denotes the reacting species and k is the rate constant. Besides the reaction scheme, of course, we also need to consider the transport mechanism responsible for bringing the particles in contact. Although many cases have been treated explicitly in the literature[1,3], we shall confine ourselves exclusively Brownian diffusion.

Another, apparently closely related reactions the following:

$$A + B \xrightarrow{k} \text{inert} \qquad (2)$$

This reaction has been used extensively to model various recombination phenomena in semiconductors[4-6] as well as annihilation of magnetic monopoles in the early universe[7] In both of these cases, we shall be interested in the question as to whether the traditional rate equation approach describes those systems correctly. As we shall see, the two systems mentioned above show radically different behavior in this respect, even though the rate equations for both systems are virtually identical. There is, therefore, an obvious independent interest in understanding the differences betweeen these two systems.

Before we go into more specific details, some very general facts should be pointed out concerning these reaction schemes: firstly, they are strictly irreversible, and hence do not satisfy any kind of detailed balance condition. This means that any system well described by such a scheme must be very far from equilibrium indeed, so that thermodynamic concepts do not find application here. Another indication of this is the fact that there are no non-trivial equilibrium states to either of these reaction schemes. In a sense, these may be said to be among the simplest nontrivial reaction schemes. Many far more elaborate models —such as the Brusselator or Schlögl's first and second models— have been extensively studied.[8-10] Typically, these models do have non-trivial equilibrium states but do not satisfy detailed balance, so that the equilibrium states are not the Gibbs states of any hamiltonian. The behavior of such models is highly complex and they shall not concern us directly here. It should, however, be noted that the two models introduced above already contain a part of this complexity and that the tools developed to derive exact results for these may also find application in their more complex counterpart.

As a final remark, let us note the following: as noted above, to specify a reaction fully, both the reaction scheme and the transport mechanism must be given. This therefore introduces two time scales in the problem: a time t_R for the reaction and another t_D for the diffusive transport. In the following, we shall examine both the limiting cases $t_R \ll t_D$ and $t_R \gg t_D$. The latter is known as the reaction-limited case and it is well-known[11] that the rate equations in this case describe the physics correctly. The former case, known as the diffusion limited case, can display marked deviations from classical behavior, which shall be described in the following. In section 2 we will describe the mean-field theory for the coalescence reaction $A + A \to A$, as well as numerical simulations in space dimensions d of one, two and three. It will appear that the salient characteristics of mean-field behavior are shown in the three dimensional case, but not in any of the other cases. In section 3, we will repeat this analysis for the case of the recombination reaction $A + B \to$ inert, showing that the rate equations predict identical behavior to the other reaction, whereas the numerical results clearly show the contrary. In section 4, we will attempt to develop a rudimentary theory to account for the observed discrepancies. Further, we will apply it to another well-studied problem,[12-14] namely the study of the effect of a source term on either of those reactions.

2. THE COALESCENCE REACTION: RATE EQUATION AND NUMERICAL RESULTS

Let us first define the lattice model introduced by Kang and Redner[21] that we will be using to simulate this reaction, so as to have a precisely defined procedure describing what we mean by each of the parameters involved. Note first that since we are interested in long-time —and hence low concentration— properties, the lattice approximation is most likely to be very reasonable. We now assume each of the points of the lattice to be either occupied by an A particle or else empty. At each time step, a point is chosen at random and, if it contains a particle, this particle is moved along each axis by a number of lattice lengths randomly chosen between $-\ell$ and ℓ, where the parameter ℓ controls the diffusion of A.

Let us now first consider the coalescence reaction in the most extremely reaction limited case: let us assume that a particle of A can move in one time step to any point in space whatsoever, i.e., that ℓ is chosen equal to L. Under these circumstances, there is no influence of the position of the various particles on the process, so that the following master equation can be derived for the probability $P_N(t)$ of having N particles at time

t in the system:

$$\frac{\partial P_N(t)}{\partial t} = k\left(\frac{N(N+1)}{V}P_{N+1}(t) - \frac{N(N-1)}{V}P_N(t)\right) \qquad (3)$$

where $V = L^d$ is the volume of the system. This can be analyzed in the limit of large V, using the well-known Ω-expansion of van Kampen[11] with the result that, for the concentration of particles c, the following equation holds:

$$\frac{dc}{dt} = -kc^2 \qquad (4)$$

with the solution

$$c(t) = \frac{c(0)}{1 + kc(0)t} \sim (kt)^{-1} \qquad (t \to \infty) \qquad (5)$$

This is, of course, nothing else than the ordinary rate equation. The object of this remark is to show that the model described above does yield, as a special case, the rate equation as an *exact* result.

A corresponding result is also readily derived for the fluctuations of the concentration, which will always differ from zero at finite volume. Define:

$$\sigma(t) = \frac{\langle N^2 \rangle - \langle N \rangle^2}{V} \qquad (6)$$

For this quantity, one obtains using the same methods:

$$\frac{d\sigma(t)}{dt} = kc(t)^2 - 4kc(t)\sigma \qquad (7)$$

with the result:

$$\sigma = \frac{c}{3} + \text{const.} \cdot c^4 \sim \frac{c}{3} \qquad (t \to \infty) \qquad (8)$$

A comparison between these exact results and simulations of the model described above for $\ell = L$ are displayed in Figures 1 and 2. The agreement is remarkably good. It should be noted in particular that excellent agreement exists even at very small times, not only in the asymptotic behavior. This is therefore a fairly stringent test of the fact that what we are simulating is indeed correctly described by equation (3) in the limit where $\ell = L$.

We note two salient features of the mean-field solution: firstly, the solution has a t^{-1} dependence, secondly, solutions to various values of k should asymptotically become identical as a function of the dimensionless time $\tau = kt$. This should still be the case for solutions with various values of the diffusion constant, since this does not enter in the equations at all. In fact, it can be seen that this remains the case if a diffusion term is added to the rate equation (4).

The actual results for d equal to one, two and three respectively are shown in Figures 3–6. For $d = 1$, it is manifest that mean-field theory is wholly inadequate: the concentration shows, to great accuracy, a $t^{-1/2}$ behavior and solutions to different values of k do not fall on the same curve as a function of $\tau = kt$, whereas solutions to different values of $D = \ell(\ell+1)/2$ fall on the same curve as a function of $\tau = Dt$. This state of affairs can

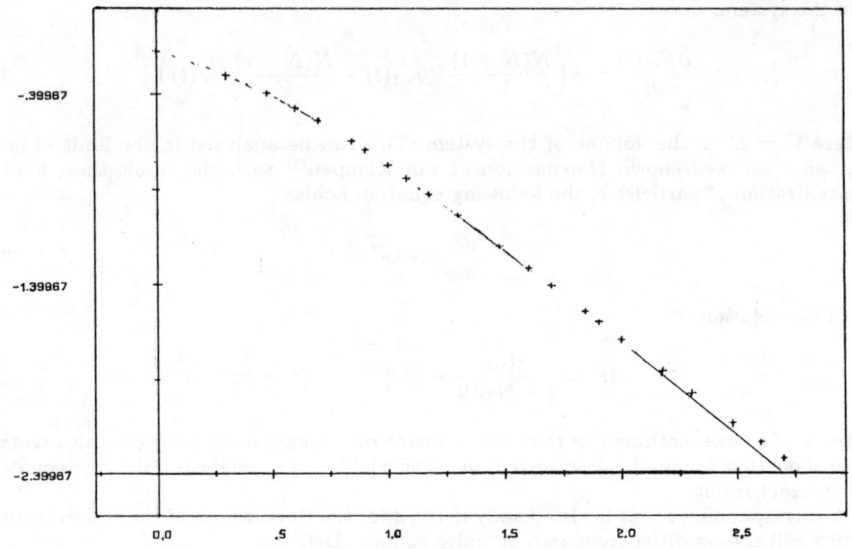

FIGURE 1. Plot of $\log c(t)$ against $\log t$. The solid line represents the theoretical prediction, whereas the plus signs represent simulation data. Note the agreement up to the earliest times.

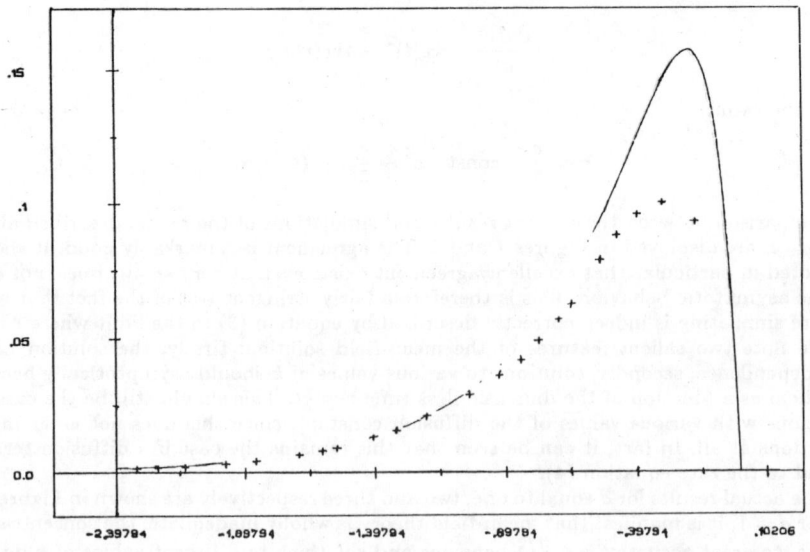

FIGURE 2. Plot of $\sigma(t)$ versus $\log c(t)$. The Solid line represents the theory and the plus signs simulation results. The reason for the discrepancy at early times (large c) is unclear.

Coalescence in one dimension

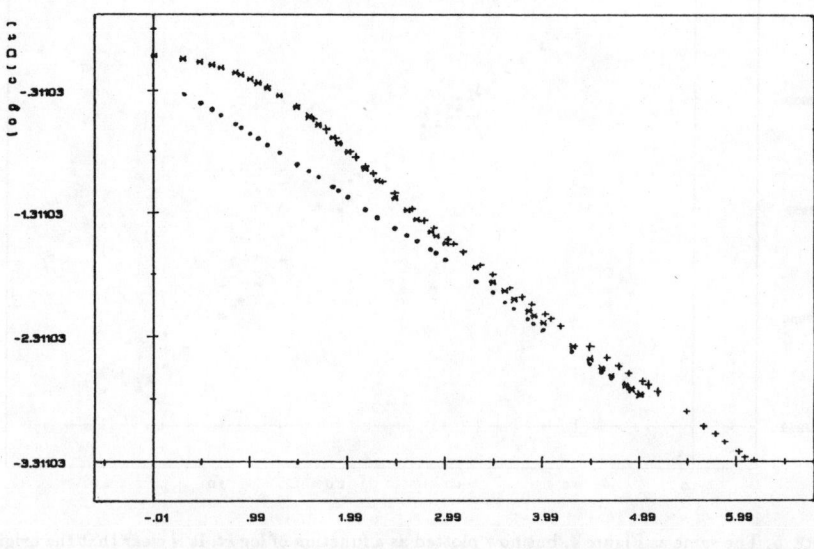

FIGURE 3. Plot of $\log c(t)$ against $\log Dt$ in three simulations of the coalescence model in one dimension. The points represent the case $D = 1$ and $k = 1$, the plus signs the case $D = 15$ and $k = 1$ and the stars the case $D = 1$ and $k = 0.1$. Note how all three converge in the asymptotic regime. The transients observed for low k or high D correspond to mean-field behavior.

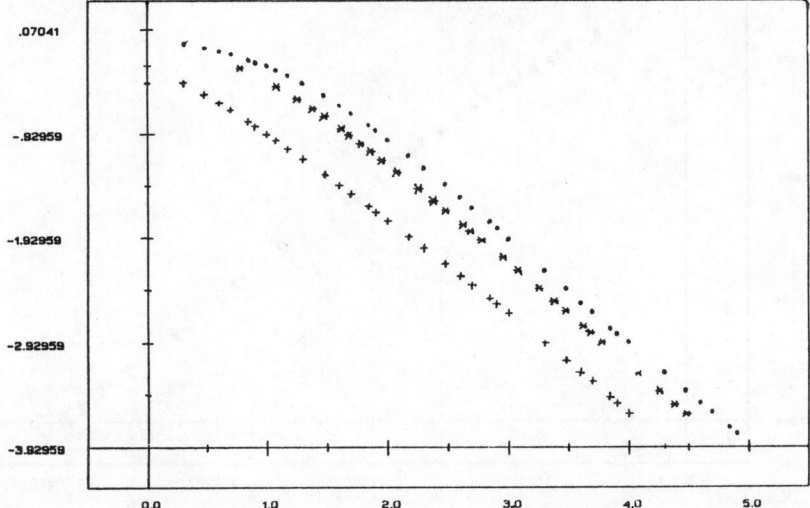

FIGURE 4. Plot of $\log c(t)$ against $\log Dt$ in three simulations of the coalescence models in two dimensions. It is clear that the three curves do not coincide, but the plus signs and the stars appear to move closer to each other, so that the asymptotic behavior is unclear. The plus signs correspond to $D = 1$ and $k = 1$, the stars to $D = 6$ and $k = 1$ and the points to $D + 1$ and $k = 0.1$.

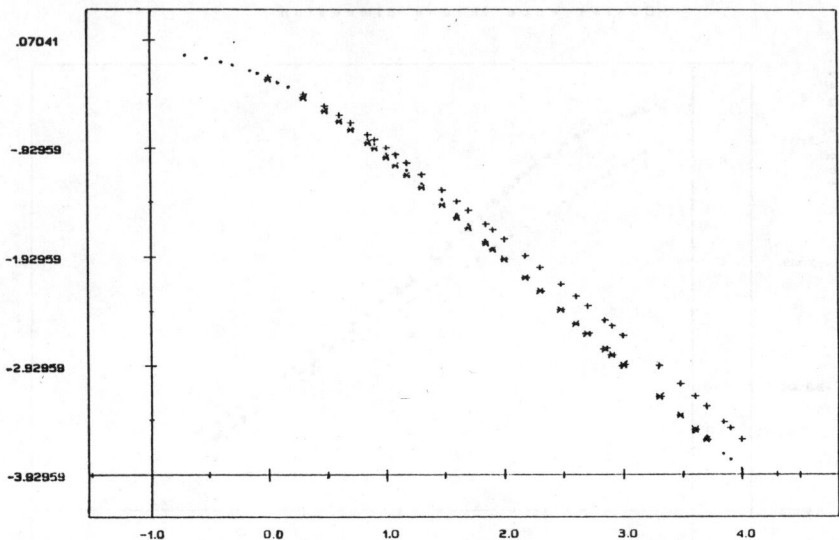

FIGURE 5. The same as Figure 4, but now plotted as a function of log kt. It is clear that the original agreement is lost as time goes on. The coincidence between the star and the points presumably means that both the low reaction rate and the high diffusion constant were enough to ensure mean-field behavior over the whole range of time studied.

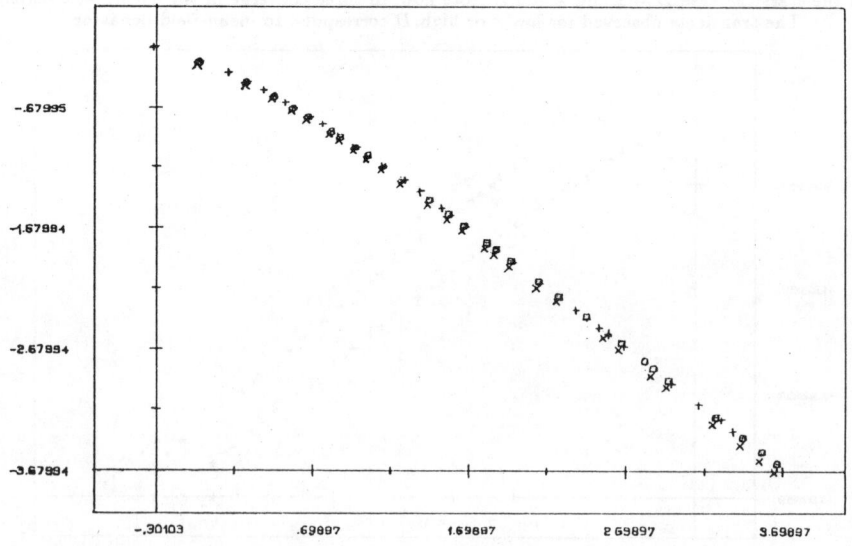

FIGURE 6. Plot of $\log c(t)$ against $\log kt$ for three simulations of the coalescence reaction in three dimen- sions. The stars correspond to $D = 1$ and $k = 1$, the crosses to $D = 2$ and $k = 1$ and the plus signs to $D = 1$ and $k = 0.5$.

be summarized by saying:

$$c(t) \sim (Dt)^{-1/2} \qquad (t \to \infty) \qquad (9)$$

As is seen quite clearly in the comparison of $k = 1$ and $k = 0.1$, the asymptotic behavior of $c(t)$ does not depend on k in any way. Indeed, it is true that for some time the mean-field solution seems to fit in the latter case. This is not surprising, since a low reaction rate mimics the conditions that we have seen to lead to the rate equation exactly. This has an obvious parallel in the theory of critical phenomena, where the effect of fairly long-range interaction is to introduce a fairly large region where mean-field exponents can be observed. Nevertheless, in both cases, the anomalous behavior is the one that eventually dominates. In our case, note that the amplitude of the asymptotic behavior is not affected by k but only by D.

In two dimensions, as shown in Figures 4 and 5, the situation is far less clear cut: a straightforward measurement of the time-dependence of $c(t)$ yields approximately a t^{-1} behavior, in good agreement with the rate equations. However, if solutions to various values of k are plotted as a function of kt, it is found that the solutions systematically diverge from each other as time increases. If, on the other hand, they are plotted as a function of Dt, the initial discrepancy decreases very slightly as $t \to \infty$, though it certainly has not disappeared even at the largest accessible times. This very strongly suggests that we are at the upper critical dimension and are observing logarithmic corrections. These conjectures will later be shown to have a theoretical foundation.

The behavior in three dimensions is displayed in Figure 6. It is seen that all the features mentioned above are indeed found, so that it appears justified to say that rate equations describe the coalescence reaction correctly in three dimensions. A few words of caution are nevertheless in order: first, it should be noted that $c(t)$ goes as $(k't)^{-1}$ where k' is a number proportional but in general not equal to the reaction probability k. Further, the proportionality constant depends on the lattice used as well as showing a weak dependence on D, going to one as $D \to \infty$, as is necessary to obtain the right limit in the case first described, where jumps can be arbitrarily large.

3. THE RECOMBINATION REACTION: RATE EQUATIONS AND NUMERICAL RESULTS

The description of the model for the recombination reaction is exactly identical to the one for the coalescence reaction, except that we now have two different species present and that no reaction ever occurs between two particles of the same species. This was discussed by Kang and Redner[22] together with extensive numerical simulations. In the following, we will always specialize ourselves to the case where the initial number of A particles is exactly equal to the number of B particles. The other case is similar to another problem[15] known as the scavenger reaction, the behavior of which is quite different.

Following the same procedure as in the last section, rate equations for the reaction can be derived in a straightforward way: if $c_{A(B)}(t)$ is the concentration of the A- and B-species respectively, one obtains:

$$\dot{c}_A = -kc_Ac_B$$
$$\dot{c}_B = -kc_Ac_B \qquad (10)$$

Since $c_A(0) = c_B(0)$, it follows from the constancy of $c_A - c_B$ that $c_A(t) = c_B(t)$ for all t. From this it follows that, in this case, the rate equations for the coalescence and recombination reactions are identical. Note, however, that the identity of the initial concentrations was essential to derive this result, since otherwise we would have exponential

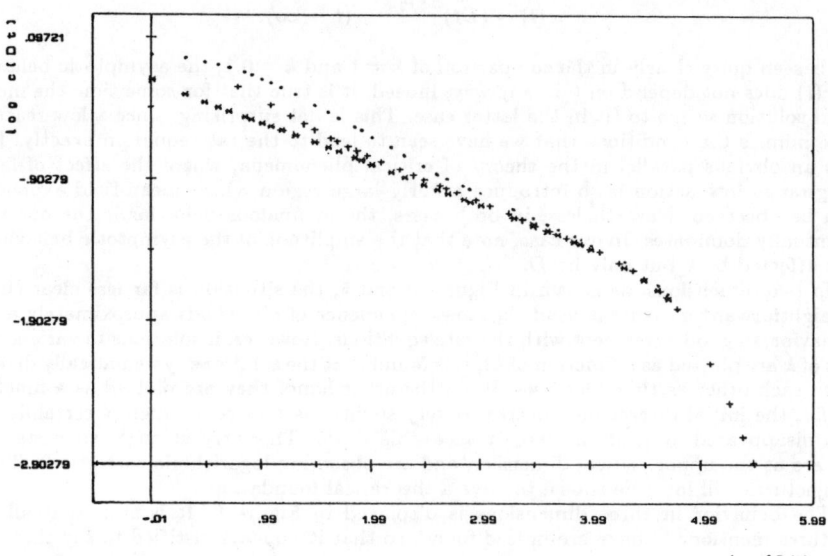

FIGURE 7. Plot of $\log c(t)$ against $\log Dt$ for three simulations of the recombination reaction in one dimension. The stars correpond to $D = 1$ and $k = 1$, the plus signs to $k = 1$ and $D = 6$ and the points to $D = 1$ and $k = 0.1$.

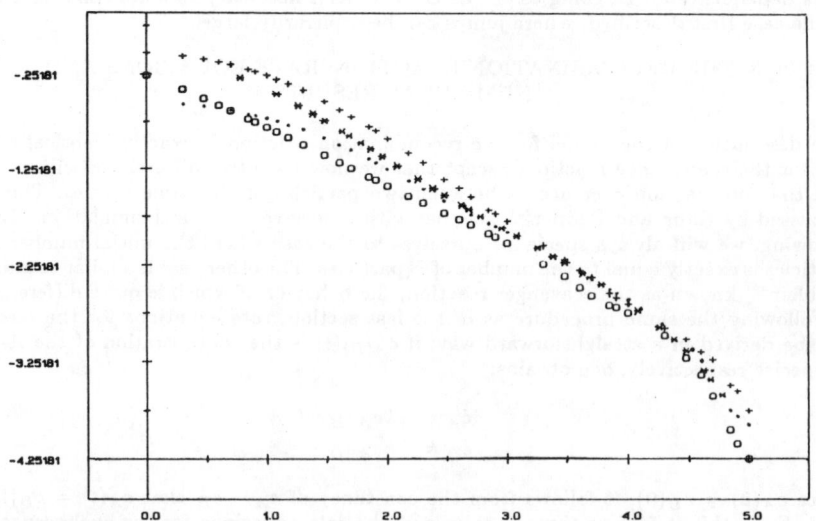

FIGURE 8. Plot of $\log c(t)/\sqrt{c(0)}$ against $\log Dt$ for four simulations of the recombination reaction in two dimensions. The open squares correspond to $D = 1$ and $k = 1$, the stars to $D = 6$ and $k = 1$ and the plus signs to $D = 1$ and $k = 0.1$. The points correspond to $D = k = 1$ but with an initial concentration of $c_0 = 0.1$.

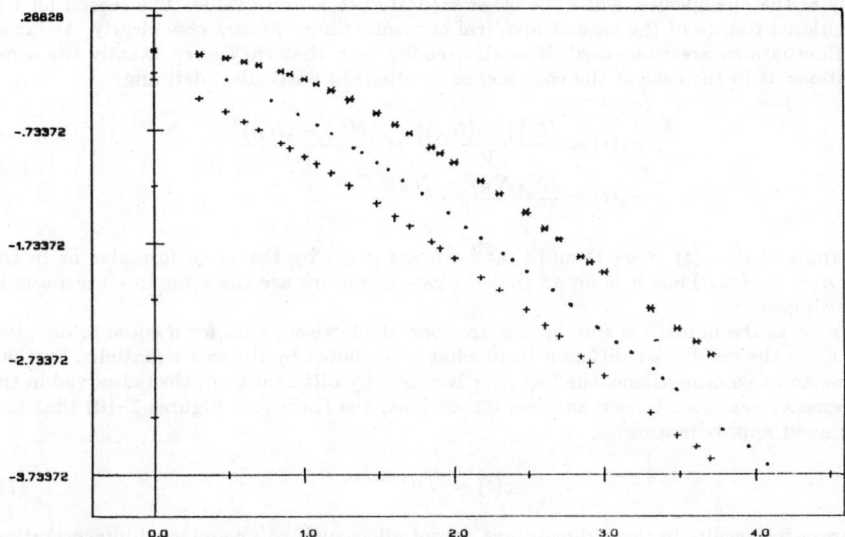

FIGURE 9. Plot of $\log c(t)$ against $\log t$ for three simulations of the recombination reaction in three dimensions. The plus signs correspond to $D = 1$ and $k = 1$, the points to $D = 3$ and $k = 1$ and the stars to $D = 1$ and $k = 0.1$.

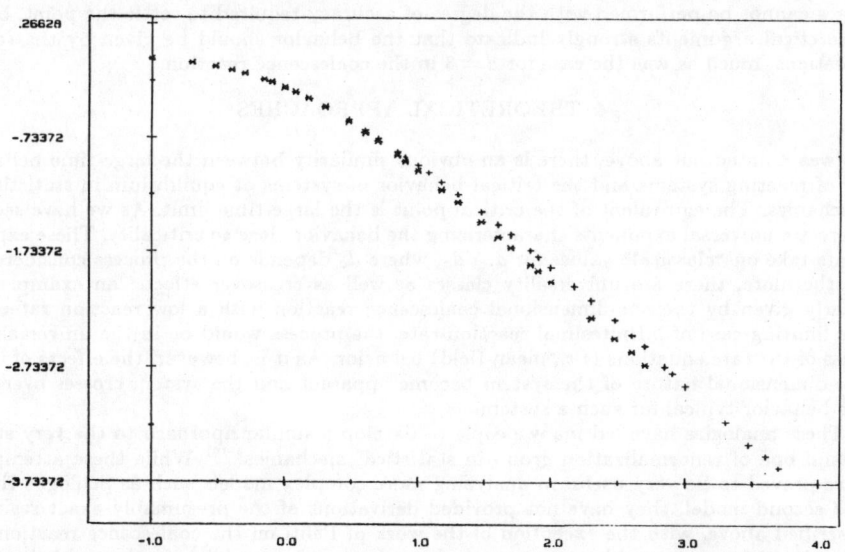

FIGURE 10. Same as Figure 9 but plotted against $\log kt$. The same remarks apply generally as to Figures 4 and 5.

decay of the one species, while the other saturated to a finite value. The reason for the exceptional nature of the case of identical concentrations appears now clearly. As far as the fluctuations are concerned, it is also readily seen that they obey exactly the same equations as in the case of the coalescence reaction. In particular, defining

$$\sigma_1(t) = \frac{\langle N_A^2 \rangle - \langle N_A \rangle^2}{V} = \frac{\langle N_B^2 \rangle - \langle N_B \rangle^2}{V}$$
$$\sigma_2(t) = \frac{\langle N_A N_B \rangle - \langle N_A \rangle \langle N_B \rangle}{V} \tag{11}$$

one finds that $\sigma_1(t) = \sigma_2(t)$ and that both are given by the same formulae as in the $A + A \to A$ case. Thus it is found that the rate equations are the same in all respects in the two cases.

As far as the numerical simulations are concerned, we see that for d equal to one, two and three the results are different from what is predicted by the rate equations. Further, in one and two dimensions, the behavior is markedly different from that observed in the coalescence reaction. In one and two dimensions, one finds (see Figures 7–10) that to a very good approximation

$$c(t) \sim (Dt)^{-d/4} \tag{12}$$

whereas the results in three dimensions do not allow such an unequivocal interpretation. However, it is readily seen that, also in this case, the solutions to various values of k do not fall on one curve as a function of kt. This is a clear indication that we still are in an anomalous regime. As we shall see, theoretical considerations[7] strongly suggest that equation (12) holds for all $d \leq 4$. For obvious reasons, simulations at dimensions $d \geq 4$ cannot be performed with the degree of accuracy required to settle the point, but theoretical arguments strongly indicate that the behavior should be given by the rate equations, much as was the case for $d = 3$ in the coalescence reaction.

4. THEORETICAL APPROACHES

As was pointed out above, there is an obvious similarity between the large-time behavior of reacting systems and the critical behavior of systems at equilibrium in statistical mechanics. The equivalent of the critical point is the large-time limit. As we have seen, there are universal exponents characterizing the behavior close to criticality. These exponents take on "classical" values for $d \geq d_c$, where d_c depends on the process considered. Furthermore, there are universality classes as well as crossover effects: an example is clearly given by the one-dimensional coalescence reaction with a low reaction rate: in the limiting case of infinitesimal reaction rate, the process would be in the universality class of the rate equations (i.e., mean-field) behavior. As it is, however, the effects of the one-dimensional nature of the system become apparent and the system crosses over to the behavior typical for such a system.

These analogies have led many people to develop a similar approach to the very successful one of renormalization group in statistical mechanics.[8-10] While these attempts have proved to be very useful in analyzing more complex models such as Schlögl's first and second model, they have not provided derivations of the presumably exact results described above, with the exception of the work of Peliti on the coalescence reaction.[17] They have, however, yielded a very appealing picture of the origin of universal behavior in such systems, as well as of crossover and related phenomena. On the other hand, in the mathematical literature, similar models have been analyzed using an analysis of the ensemble of histories of the system.[18] This has yielded very detailed exact results, but no clear picture of the origins of universal behavior. Thus, the behavior of the one dimensional coalescence model can be obtained exactly for the case $k = 1$, but the independence of the asymptotic behavior from k does not follow.

On the other hand, many heuristic arguments have been advanced to explain the behavior described above. They are appealing, due to their simplicity. What is more, they are the only arguments for the recombination reaction that I am aware of. On the other hand, their drawback is that they are not of a systematic nature, but that they require the identification of the relevant elements. In the following, I will present these arguments as well as a very simplified form of "real space renormalization group" on the ensemble of all histories of the system.

The basic heuristic arguments describing the behavior of both the coalescence and the recombination reaction are found in Toussaint and Wilczek[7] as well as in Zumofen et al.[16] The argument for the coalescence reaction goes as follows: any A particle visits on the average $S(t)$ different sites in a time t, where the asymptotic behavior of $S(t)$ depends on dimension in the following manner:

$$S(t) \sim \sqrt{Dt} \qquad (d = 1)$$
$$S(t) \sim \frac{Dt}{\ln Dt} \qquad (d = 2) \qquad (13)$$
$$S(t) \sim \text{const.} \cdot t \qquad (d > 2)$$

where the constant in the last equation depends weakly on D and goes to one as D goes to infinity. This behavior is clearly very reminiscent of the one observed for the concentration of A particles in the coalescence model. This can be understood as follows: considering any A particle during a period of time t, it will have visited on the order of $S(t)$ different sites and will at best have reacted with one particle on each site, leaving a depleted region of $S(t)$ sites. Assuming a similar development everywhere, we see that we are left with no less than one particle for every $S(t)$ sites, i.e.,

$$c(t) \sim \frac{1}{S(t)} \qquad (14)$$

Strictly speaking, the above reasoning only gives a bound on $c(t)$. As we have seen in the numerical data, however, this bound appears to be remarkably good. This can be understood as follows: it is well-known that a random walk eventually visits every site of the lattice if $d \leq 2$. Therefore, the picture developed above is reasonable for $d \leq 2$, since every site is visited a large number of times, so that the region visited is indeed eventually emptied, no matter what the reaction rate is. This explains the fact that the essential time scale in these problems was given by D and not by k. On the other hand, in three dimensions and above, every site is visited on the average a certain (non-divergent) number of times. For this reason, the reaction rate becomes relevant. We thus see that the result of this approximate reasoning is identical to the prediction of the rate equations.

For the case of the recombination reaction, the situation may at first appear similar. There is, however, a crucial difference: the quantity $c_A(t) - c_B(t)$ is conserved. This means that an excess of either type of particle inside a given region can only decay by diffusion through the boundaries of the region. The formation of large A-rich or B-rich domains can therefore only be prevented by diffusion. On the other hand, such domains would considerably interfere with the reaction, since only the particles at the domain boundaries could react. To investigate this, let us consider a part of the system of size

L'. Originally, the number of (say) A particles is

$$N_A = c(0)(L')^d + \text{const.} \cdot \sqrt{c(0)} \, (L')^{d/2} \tag{15}$$

where $c(0)$ is the initial concentration of either A or B and the second term represents an estimate of the expected particle number fluctuations. If we assume that this subsystem is, in fact, closed, so that the difference in the numbers of the two species cannot be eliminated, we would find that within a time of the order of $(L')^{d/2}$ the dominant species would have reacted completely with the minority species, leaving a concentration of the order of $(L')^{-d/2}$ of the majority species. But, as long as $d < 4$, this time is much shorter than $(L')^2$, which is the time necessary for diffusion to take effect and eliminate the excess particles. Thus, for $d < 4$, the system forms ever growing domains in which one of the two substances predominates. To determine the behavior of $c(t)$ out of these remarks, one can proceed as follows: at $t = 0$, the number of A particles is given by $c(0)(L')^d$. After a time $t \approx (L')^2/D$, the number remaining is $\sqrt{c(0)} \, (L')^{d/2}$. Eliminating L' out of these three relations, we obtain

$$c(t) \sim \sqrt{c(0)} \, (Dt)^{-d/4} \tag{16}$$

as was observed in the numerical part.

Summarizing, we note that the cause for the discrepancy lies in the neglect of spatial correlations induced by the combined effects of reaction and diffusion. In fact, it is easy to write down an exact equation for $c(t)$

$$\dot{c}(t) = -k g^{(2)}(0) \tag{17}$$

where $g^{(2)}(0)$ is the probability of finding two particles (of different species in the recombination case) at zero separation one from the other. The problem, of course, then reduces to finding a reasonable expression for this function. There is, of course, an exact relation between it and an appropriate three-point correlation function. This leads to very complicated equations, even if only the very simplest decoupling schemes are used. On the other hand, it provides some insight into the nature of the rate equations approximation: they are simply a zero order decoupling scheme, replacing $g^{(2)}(0)$ by its limiting value $g^{(2)}(\infty)$. This now shows the reason for the failure of rate equations in low dimensions: due to the reaction, spatial correlations build up around the particle (in particular, the particles capable of reacting with the particle under consideration will be depleted) but diffusion in low dimensions will not be sufficient to destroy them. In high enough dimensions, however, diffusion is powerful enough to carry them away, at least to such an extent that the "correlation hole" around each particle only has the effect of changing the effective value of the reaction rate k.

Another approach, that brings out more strongly the similarities to the renormalization group, is the following: the reaction, instead of being considered as a strictly dynamical phenomenon, is viewed as an ensemble of histories, i.e, of the trajectories of all the particles visualized on a space-time diagram. Clearly, there is no easy way to determine the probability of a given history. But we do not require such detailed information: rather, it is entirely sufficient to know how histories are changed under scale transformations. To this end, let us start by considering a system of finite extent, which is allowed to run until the reaction has been completed. If the dimension of the original system was L, then in order to scale the system by a factor b, it is sufficient to stretch all spatial lengths by a factor b and time by a factor b^α such that the reaction also goes to completion in the renormalized time interval. The determination of α is then the only problem that remains to be solved and this can, as we shall see, be done straightforwardly for the reactions considered above.

For the coalescence reaction, the reaction has run to completion when only one particle remains in the system. If one therefore starts with N particles, after a time t_N there will only be one particle left. If we now scale the system by a factor b in all its lengths, we have b^d different systems evolving over a time $b^\alpha t_N$. This has the following consequences for the behavior of the concentration of A particles after the reaction has been completed

$$c' = b^{-d}c \quad (18)$$

where the prime denote the renormalized quantities. From this follows that c scales as $t^{-d/\alpha}$. Note that the behavior we have observed numerically strongly suggests $\alpha = \max(d, 2)$ Let us now consider the possible values of α. Note that for the diffusion constant D one has

$$D' = b^{\alpha-2}D \quad (19)$$

so that diffusion eventually becomes infinitely rapid as the system is renormalized if α is chosen to be greater than two. In this case, however, one knows that mean-field theory is applicable, so that $c(t) \sim (kt)^{-1}$. This in turn implies that the only possible choice for α in this case is d. On the other hand, if we chose α less than two, the diffusion will eventually become infinitely slow, thus precluding any possibility of the reaction going to completion. The only two valid choices for α are therefore two and d if $d \geq 2$. It is not hard to show that the reaction will run to completion if $\alpha = \max(d, 2)$.

To do this, we estimate the likelihood that a given particle reacts in a given time interval $[t, t+\tau]$ and how this likelihood scales if lengths are scaled by a factor b and the times by a factor b^2. Let us define n as the number of particles accessible to the particle under consideration in a time τ. Further, let us call W the total space-time volume accessible to the particle in time τ. The probability of reaction is then given by

$$p \propto k\frac{n\tau^2}{W} \quad (20)$$

If we now rescale lengths and times as above, we obtain, using the fact that the number of particles coming out of the reaction does not depend on b

$$W' = b^{d+2}W$$
$$n' = n \quad (21)$$
$$p' = b^{2-d}p$$

Therefore, the reaction probability increases as the system is renormalized if $d \leq 2$ and decreases otherwise. This means that the reaction does indeed proceed to completion if $d \leq 2$ and α is chosen to be two. Of course, the increase of the reaction probability does not proceed indefinitely. Rather, it stops as the probability reaches one, so that the parameter p, in which the dependence on the reaction rate k appears, is irrelevant in low dimensions. This is in good agreement with our numerical observations. In high dimensions, on the other hand, it is easy to see that the diffusion constant eventually becomes irrelevant and the reaction constant k sets the time scale.

For the recombination reaction, the crucial difference to notice is that, as the reaction runs to completion in a finite system, the number of particles remaining is of the order of $\sqrt{N}$ where N is the initial number of particles. This means that if we rescale lengths by b and time by b^α, we obtain

$$c' = b^{-d/2}c \qquad (22)$$

and hence c scales as $t^{-d/2\alpha}$. Thus, if we chose $\alpha > 2$, we would obtain, by the same reasoning as above, $c(t) \sim (kt)^{-1}$, so that the only possible such value for α is $d/2$. Again, however, this choice only becomes possible when $d \geq d_c = 4$. A very similar consideration to the previous one shows that the probability p of a particle reacting indeed increases indefinitely as the system is renormalized if $d < 4$. This is seen as follows: using the same notation as previously, we obtain the following scaling behavior

$$W' = b^{d+2}W$$
$$n' = b^{d/2}n \qquad (23)$$
$$p' = b^{2-d/2}p$$

where the second equation reflects the fact that, upon increasing the number of particles initially in a system, the number of resulting particles increases as the square root of that number.

Some facts should be noted about the preceeding analysis of the recombination reaction. First, it does not require strict conservation of $c_A(t) - c_B(t)$. Thus the following reaction scheme studied in reference 20 also has the same behavior

$$A + B \longrightarrow A \quad \text{or} \quad B \qquad (24)$$

This is readily understood by noting that the finite system has the same behavior, i.e., that the reaction usually completes with on the order of $\sqrt{N}$ particles remaining. On the other hand, the following system is in the same universality class as the coalescence reaction

$$A + B \xrightarrow{K} \text{inert}$$
$$A + A \xrightarrow{k} B + B \qquad (25)$$
$$B + B \xrightarrow{k} A + A$$

since the reaction always ends up with one or two particles. This agrees with numerical results and shows that it does not matter so much whether the conservation law is exact or approximate, but rather on the specific ways in which the reaction can terminate.

Further, the approach can be generalized to more complex systems[19] such as

$$A + B \longrightarrow \text{inert}$$
$$B + C \longrightarrow \text{inert} \qquad (26)$$
$$A + C \longrightarrow \text{inert}$$

In this case, it is not trivial to find how many particles will remain on the average at the end of the reaction if there were N particles originally. In a mean-field approximation, however, it can be done (see reference 19 for details). The result is $N^{1/4}$, leading to the results obtained in reference 19, *i.e.*

$$c(t) \sim t^{-3d/8} \tag{27}$$

as long as $d \leq 8/3$.

In a similar way, the following system with a backward reaction can be investigated

$$A + A \xrightarrow{k} A$$
$$A \xrightarrow{R} A + A$$

Indeed, if one considers a cell containing on the average one particle and waits a time in which, on the average, one reaction will take place, it is again possible to scale lengths by b and times by b^2. To maintain the conditions described above, it is necessary to rescale R and the equilibrium concentration c_0 as follows

$$\begin{aligned} R' &= b^{-2}R \\ c_0' &= b^{-d}c_0 \end{aligned} \tag{28}$$

so that c_0 scales as $R^{-d/2}$. Again, this argument can be seen to break down when $d > 2$. An argument entirely similar to the one performed for the coalescence reaction shows that in this case c_0 goes as R, as predicted by the rate equation

$$\dot{c} = -kc^2 + Rc \tag{29}$$

These results have been discussed in a slightly different way by Anacker *et al.*[12, 13] with identical results.

5. CONCLUSIONS

Summarizing, we have discussed two very elementary types of binary reactions. We have seen that it is necessary to distinguish between the case where the rate limiting step is the reaction and the opposite case, where it is the transport mechanism (in our case, diffusion). As we have seen, in the former case, a very efficient transport mechanism can be assumed to destroy any spatial correlations that might have built up. This means that the probability of an encounter between two particles can be reliably estimated by c^2, as is usually done in deriving rate equations.

On the other hand, in the latter case, the spatial correlations created by the simultaneous operation of the reaction and the diffusion processes can build up if the space dimension is sufficiently low. In this case, as we have seen, the actual rate of reaction does not set the time scale any more. Rather, the diffusion constant does, as the effects due to correlation eventually overwhelm the behavior expected from the rate equations. We have seen how various heuristic arguments can be developed in order to predict the various types of behavior occurring in different systems. In particular, we have seen that a certain universality holds, so that changing only "irrelevant" details in the reaction scheme (such as the reaction rate or changing from coalescence to one-species annihilation) does not affect the behavior in any way, except in some cases over a large but finite crossover regime. It would clearly be of considerable interest to extend these approaches to the case of more complex reactions, in particular such as have non-trivial equilibria. It might also be of interest to generalize the above remarks to the spatially inhomogeneous case where diffusion can be expected to play a much more complex role.

REFERENCES

1. Drake, R.L., *Topics in Current Aerosol Research* 3 (eds. G.M. Hidy and J.R. Brock).
2. Smoluchowski, M.v., *Phys. Zeitschr.* 17, 593 (1916).
3. Friedländer, S.K., *Smoke, Dust and Haze*, Wiley, New-York (1977).
4. Scher, H. and Montroll, E.W., *Phys. Rev. B* 12, 2445 (1975).
5. Shlesinger, M.F., *J. Chem. Phys.* 20, 4813 (1979).
6. Ngai, K.L. and Lin, F.S., *Phys. Rev. B* 26, 1049 (1981).
7. Toussaint, D. and Wilczek, F., *J. Chem. Phys.* 78, 2642 (1983).
8. Doi, M., *J. Phys. A: Math. Gen.* 9, 1465 (1976).
9. Grassberger, P. and Scheunert, M., *Fortschr. Phys.* 28, 547 (1980).
10. Peliti, L., *J. Physique* 46, 1469 (1985).
11. van Kampen, N.G., *Stochastic Processes in Physics and Chemistry* (North-Holland, Amsterdam, 1981).
12. Anacker, L.W. and Kopelman, R., *J. Chem. Phys.* 81, 6402 (1984).
13. Anacker, L.W., Parson, R.P. and Kopelman, R., *J. Phys. Chem.* 89, 4758 (1985).
14. ben-Avraham, D. and Doering, C.R. (preprint).
15. Redner, S. and Kang, K., *J. Phys. A: Math. Gen.* 17, L451 (1984).
16. Zumofen, G., Blumen, A. and Klafter, J., *J. Chem. Phys.* 82, 3198 (1985).
17. Peliti, L., *J. Phys. A: Math. Gen.* 19, 973 (1986).
18. Torney, D.C. and McConnell, H.M., *J. Phys. Chem.* 87, 1441 (1983); *Proc. Roy. Soc. London Ser. A* 387, 147 (1983).
19. ben-Avraham, D., *J. Stat. Phys.* 48, 315 (1987).
20. ben-Avraham, D. and Redner, S., *Phys. Rev. A* 34, 501 (1986).
21. Kang, K. and Redner, S., *Phys. Rev. A* 30, 2833 (1984).
22. Kang, K. and Redner, S., *Phys. Rev. A* 32, 435 (1985).

SOLUBLE RANDOM-MATRIX MODEL FOR DISSIPATIVE TWO-LEVEL SYSTEMS

Pedro Pereyra

*Departamento de Ciencias Básicas,
Universidad Autónoma Metropolitana-Azcapotzalco,
02200 México, D.F., México*

ABSTRACT. For times larger than the duration of a collision and smaller than the Poincarè recurrence time and assuming a random-matrix model for the interaction between a two-level system and a thermal bath, we calculate the survival probability of still finding the system, at time t, in the same state in which it was prepared at $t = 0$.

1. INTRODUCTION

There is a large variety of physical and chemical systems whose dynamics can be studied by means of two-state models where an interaction with a given environment is considered.[1-7] Although a complete solution of these problems seems rather difficult, there is an increasing interest for studying the so called "Dissipative two-level systems"; important progress has been achieved. In this contribution we present an alternative and soluble model for the description of these systems. It is known that some of the real-life systems are "intrinsically two-state systems" because they possess a discrete degree of freedom that can take only two values (spin 1/2 particles, polarization of a photon, etc.). Other systems possess a continuous degree of freedom q whose dynamics can be described by using a potential energy function $V(q)$ with two separate minima (paraelectric resonance relaxation, stabilization of handed molecules, molecular polarons, tunneling, etc.).

For a realistic description of the two-level systems, an interaction with their environment (heat bath, radiation field, etc.) must be taken into account. This fact introduces some complexity into the problem; but it turns out that the time evolution of the two-level systems is substantially modified by such interaction. Therefore, it is usual to consider Hamiltonians with the following structure

$$H = H_S + H_E + H_I, \tag{1}$$

where H_S describes the isolated two-state systems, H_E describes the environment and H_I the system-environment interaction. One of the well known Hamiltonians of this kind, that has been extensively studied in the literature, is the "spin-boson" Hamiltonian[4]

$$H = -\frac{1}{2}\hbar\Delta_0\sigma_x + \frac{1}{2}\epsilon\sigma_z + \sum_\alpha \left(\frac{1}{2}m_\alpha\omega_\alpha x_\alpha^2 + \frac{p_\alpha^2}{2m_\alpha}\right) + \frac{1}{2}q_0\sigma_z\sum_\alpha C_\alpha x_\alpha, \tag{2}$$

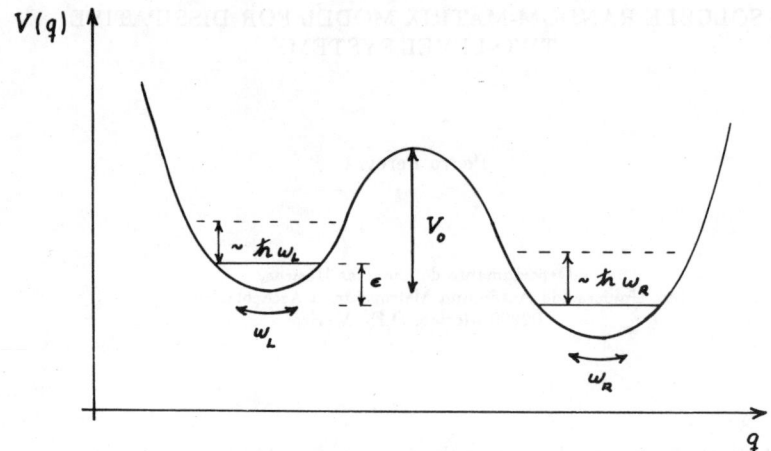

FIGURE 1. A double well potential energy function.

where the isolated two-state system is completely described by the first two terms. The set of harmonic oscillators in the third term represents the bath and the last term represents the system-bath interaction. The physical meaning of the first two terms can be found with the help of Figure 1, where a potential energy function with two minima is shown. Legget et al. discuss this case and show that the double-well system is effectively described by the Hamiltonian in equation (2) if the barrier height V_0 is large compared to $\hbar\omega_R$ and $\hbar\omega_L$ (where ω_R and ω_L are the classical small-oscillation frequencies in the right and left well separately) and the bias ("detuning") ϵ between the ground states is small compared to ω_R and ω_L. When the basis is chosen so that the eigenstate $|\psi_R\rangle$ ($|\psi_L\rangle$) of σ_z with eigenvalue $+1$ (-1) correspond to the system localized in the right (left) well, the term $(1/2)\epsilon\sigma_z$ represents the difference in the ground-state energies, while the term $(1/2)\hbar\Delta_0\sigma_x$ describes the tunneling between the wells. For many cases of practical interest $\epsilon = 0$. An alternative interpretation for the first two terms in equation (2) is that they represent a particle of spin 1/2 in the magnetic field $\mathbf{H} = -\epsilon\,\hat{\mathbf{k}} + \hbar\Delta_0\,\hat{\mathbf{i}}$. One of the quantities that has been calculated by Leggett et al. is the expectation value of σ_z as a function of t.

Another well known Hamiltonian is the "rotating wave" Hamiltonian used in quantum optics to describe the rate process of a two-state atom interacting with a radiation field, which, in a fictitious spin-1/2 representation, takes the form

$$H = \frac{1}{2}\omega_0\sigma_z + \omega_L \sum_\alpha a_\alpha^+ a_\alpha + \frac{\lambda}{2}\sigma_x \sum_\alpha (a_\alpha + a_\alpha^+). \quad (3)$$

There is no difference at all between (2) with $\epsilon = 0$ and (3) if in the last case the basis is chosen so that the eigenstates $|\varphi_R\rangle$ ($|\varphi_L\rangle$) of σ_x with eigenvalues $+1$ (-1) correspond to the system localized in the right (left) well. However, if the basis is chosen so that the eigenstates $|1\rangle$ ($|-1\rangle$) of σ_z correspond to the system localized in the right (left) well, the quantity $\langle\sigma_z\rangle_{rw}$ calculated with the rotating-wave Hamiltonian does not have the same physical meaning as $\langle\sigma_z\rangle_{sb}$ calculated using the spin-boson Hamiltonian. While $\langle\sigma_z\rangle_{sb}$ is associated with the tunneling process, $\langle\sigma_z\rangle_{rw}$ is related to the spin-flip process induced by the interaction with the bath.

In the relaxation problems, the process is frequently found to be insensitive to the details of the interaction, only a few "gross properties" being relevant for its description. This feature is not new in many-body problems, and has often been explicitly described by constructing a collection or ensemble of interactions,[7] and calculating an ensemble average of the quantity of interest: if that quantity does not vary appreciably across the ensemble, it can be reasonably represented by its average; if that were not the case, one could certainly calculate the fluctuations of the given quantity across the ensemble. This philosophy has been implemented in the study of the relaxation of a degenerate-two-level system interacting with a bath.[8] The purpose of the present paper is to extend the calculations of reference 8 to the case where one also has a Hamiltonian term associated to the two-level system. The total Hamiltonian of the problem is given as

$$H = \tfrac{1}{2}\Delta_0\sigma_z + H_B + \sigma_x V. \qquad (4)$$

The first term describes the dynamics of the isolated two-level system and can be interpreted as the energy of a spin-1/2 particle in the magnetic field $\mathbf{H} = (1/2)\Delta_0 \hat{\mathbf{k}}$. The third term represents the system-bath interaction, where V depends only on the bath variables, H_B is the bath Hamiltonian.

Suppose that at $t = 0$ we prepare the system in an eigenstate, $|1\rangle$ say, of the Pauli matrix σ_z, while the bath is in thermal equilibrium. Since $\sigma_x V$ can cause transitions between the two eigenstates of σ_z, the problem that we pose is that of calculating the probability of still finding the system in state $|1\rangle$ at time t.

We shall be able to show that the model just described can be solved *exactly*, when the time t satisfies the inequalities

$$t_{\text{coll}} \ll t \ll t_p, \qquad (5)$$

t_{coll} and t_p being times on the order of the duration of a collision (to be distinguished from the time between collisions!) and of the Poincaré recurrence time, respectively. The procedure, inspired in the one followed in reference 9 to describe relaxation phenomena in nuclear physics, essentially consists in writing the above mentioned survival probability as a series expansion in powers of the interaction, averaging term by term and then summing up the full series.

Assumptions on a phenomenological, random, time-dependent interaction, are sometimes made in relaxation studies.[10-14] We wish to remark that in the present paper such assumptions are not needed, because we work with the full, time independent Hamiltonian, and any time dependence should come out as a consequence of the model. We notice, incidentally, that the randomness assumed for such a time-dependent interaction has an entirely different origin from the one considered in the present article, since in standard statistical mechanical problems it is taken for granted that there exists *one* total Hamiltonian for the full problem, and not an ensemble of Hamiltonians as we consider here.

The paper is organized as follows. In Section 2 we define with more precision the model that was outlined above; in particular, we discuss in detail the random-matrix ensemble that we propose for the system-bath interaction. The survival probability is written in that section as a series in powers of the interaction. Some representative terms of that series are evaluated in detail in Section 3; a graphical representation for them is given, which is greatly advantageous for the evaluation of the most general term. The series can be summed, giving the final result (43) for the survival probability. Some features of the result are discussed in Section 4. Finally, Section 5 gives the conclusions of this investigation.

2. THE SURVIVAL PROBABILITY AND THE RANDOM-MATRIX MODEL

Before we discuss the assumptions on the random interaction, let us introduce the notation and write out the physical quantity that we are going to calculate. We designate by $|\alpha\rangle$, $\alpha = \pm 1$, the eigenstate of σ_z with eigenvalue α, *i.e.*

$$\sigma_z |\alpha\rangle = \alpha |\alpha\rangle, \qquad \alpha = \pm 1. \tag{6}$$

In addition, we denote by $|a\rangle$ a complete set of eigenstates of the bath Hamiltonian with eigenvalue ϵ_a, *i.e.*

$$H_B |a\rangle = \epsilon_a |a\rangle. \tag{7}$$

The states $|a\alpha\rangle$ form a complete set of states for the system-bath combination which we assume is governed by the Hamiltonian

$$H = \tfrac{1}{2}\Delta_0 \sigma_z + H_B + \sigma_x V. \tag{8}$$

We assume that at $t < 0$ the system is held in the state $|1\rangle$ and the bath in thermal equilibrium, described by the canonical ensemble

$$p_{a_i} = \frac{1}{Z} e^{-\beta \epsilon_{a_i}}, \tag{9}$$

where Z is the bath partition function. At $t = 0$ the system-bath interaction is switched on, inducing transitions between the two eigenstates of σ_z.

We are looking for the probability $P_{1 \to 1}(t)$ that at time $t > 0$ we still find the system in the state $|1\rangle$, regardless of the state of the bath, *i.e.*

$$P_{1 \to 1}(t) = \sum_{a_i, a} p_{a_i} \left| \langle 1 a | e^{-iHt} | 1 a_i \rangle \right|^2. \tag{10}$$

As we mentioned before, we use the interaction representation and expand the evolution operator in power series of the interaction H_I as

$$e^{-iHt} = e^{-iH_0 t} \sum_{n=0}^{\infty} (-i)^n \int_0^t dt_n \, H_I(t_n) \int_0^{t_n} dt_{n-1} \, H_I(t_{n-1}) \cdots \int_0^{t_2} dt_1 \, H_I(t_1), \tag{11}$$

where H_0 represents the first two terms of equation (4) and

$$\begin{aligned} H_I(t) &= e^{-iH_0 t} H_I e^{iH_0 t} \\ &= e^{-iH_0 t} V e^{iH_0 t} [\sigma_x \cos \Delta_0 t + \sigma_y \sin \Delta_0 t] \\ &= \sum_{b_j, b_{j+1}} |b_j\rangle e^{i\epsilon_{b_j} t} \langle b_j | V | b_{j+1} \rangle e^{-i\epsilon_{b_{j+1}} t} \langle b_{j+1}| \\ &\quad \times [\sigma_x \cos \Delta_0 t + \sigma_y \sin \Delta_0 t]. \end{aligned} \tag{12}$$

Because of σ_x and σ_y in (12) only the even order terms contribute to (11). Thus, the survival probability takes the form

$$P_{1\to 1}(t) = \sum_{a_i} p_{a_i} \sum_{p,q=0} (-)^{p+q} \int_0^t dt_{2p} \int_0^{t_{2p}} dt_{2p-1} \cdots \int_0^{t_2} dt_1 \int_0^t dt'_{2q} \cdots \int_0^{t'_2} dt'_1$$

$$\times \sum_{\substack{a,b_1,b_2,\ldots,b_{2p-1} \\ b'_1,b'_2,\ldots,b'_{2q-1}}} V_{a_i b_1} V_{b_1 b_2} \cdots V_{b_{2p-1} a} V_{ab'_{2q-1}} \cdots V_{b'_2 b'_1} V_{b'_1 a_i}$$

$$\times \exp\Big\{i[(\epsilon_{a_i} - \epsilon_{b_1})t_1 + \cdots + (\epsilon_{2p-1} - \epsilon_a)t_{2p}$$

$$+ (\epsilon_a - \epsilon_{b'_{2q-1}})t'_{2q} + \cdots + (\epsilon_{b'_1} - \epsilon_{a_i})t'_1]\Big\} \tag{13}$$

$$\times \exp\Big\{-i\Delta_0[t_1 - t_2 + t_3 - \cdots - t_{2p} + t'_{2q} - t'_{2q-1} + \cdots + t'_2 - t'_1]\Big\}$$

$$= \sum_{a_i} p_{a_i} \sum_{p,q} P_{1\to 1}^{(p,q)}(t, a_i)$$

which depends on the matrix elements of the system-bath interaction and expresses the survival probability as a sum of contributions coming from the various initial states a_i and the values taken by the variables p and q; the (p,q)-term is of order $2p + 2q$.

We shall now be more specific about the random-matrix model assumed for V. We propose a *local GOE* (Gaussian Orthogonal Ensemble),[8] an extension (used in reference 9) of the standard GOE[7] to be defined below.

The matrix elements V_{ab} of the operator V, in the basis defined by (7), are assumed to form a *real symmetric* matrix and, aside from the symmetry requirement, they are considered as *statistically independent Gaussian variables* with zero mean and covariance given by

$$\langle V_{ab} V_{cd}\rangle = v_a^2 (\delta_{ad}\delta_{bc} + \delta_{ac}\delta_{bd}) w_{\Delta_a}(\epsilon_a - \epsilon_b), \tag{14}$$

the angular brackets denoting an ensemble average. In other words, the only nonzero covariance is that of V_{ab} with itself, or with V_{ba}.

The weight factor $w_{\Delta_a}(\epsilon_a - \epsilon_b)$ is assumed to be a Lorenzian function of width Δ_a, thus indicating that the interaction V connects eigenstates of H_B within an energy interval $\sim \Delta_a$. The quantity ($\hbar = 1$)

$$t_{\text{coll}} \sim \frac{1}{\Delta_a} \tag{15}$$

is a time associated with one application of the interaction, and has been interpreted as the duration of a collision.[8,9] We shall always assume that Δ_a contains many bath levels, and we shall allow for the possibility of a slow dependence of Δ_a on the bath energy ϵ_a. A slow dependence on ϵ_a of the strength of the interaction v_a^2 may also occur.

We can thus visualize the matrix $\|V_{ab}\|$ as having appreciable elements inside a band of variable width $2\Delta_a$ along the diagonal.

We shall frequently encounter in what follows the quantity $\langle V_{ab}^2 \rangle \rho(\epsilon_b)$, where $\rho(\epsilon)$ denotes the density of bath states. Using (14) and defining the quantity

$$\Gamma_a = 4\pi v_a \rho(\epsilon_a) \tag{16}$$

we write $\langle V_{ab}^2 \rangle \rho(\epsilon_b)$ as

$$\langle V_{ab}^2 \rangle \rho(\epsilon_b) = \Gamma_a w_{\Delta_a}. \tag{17}$$

Since $w_{\Delta_a} \approx 0$ outside of an interval Δ_a and both Γ_a and Δ_a vary slowly with energy, we can use, in (17), either Γ_a or Γ_b, w_{Δ_a} or w_{Δ_b}, and Δ instead of Δ_a. This fact will be considerably useful later.

Once we have defined the random matrix model, the next task is to calculate the ensemble average of the survival probability $P(t)$ of equation (13), which can then be written as a sum over the various i, p, q contributions, as

$$\begin{aligned}\langle P_{1 \to 1}(t) \rangle &= \sum_{a_i} p_{a_i} \sum_{p,q=0}^{\infty} \langle P_{1 \to 1}^{(p,q)}(t, a_i) \rangle \\ &\equiv \sum_{a_i} p_{a_i} \langle P_{1 \to 1}^{a_i}(t) \rangle,\end{aligned} \tag{18}$$

where each contribution satisfies the relation

$$\langle P_{1 \to 1}^{(p,q)}(t, a_i) \rangle = \langle P_{1 \to 1}^{(q,p)}(t, a_i) \rangle^*. \tag{19}$$

The calculation of $\langle P_{1 \to 1}^{a_i}(t) \rangle$ will be carried out in the next section.

3. THE ENSEMBLE AVERAGE OF THE SURVIVAL PROBABILITY

Just to illustrate we calculate in detail some representative terms in the expansion (18). It will then be easy to infer the general values and apply them to the calculation of the survival probability

3.1. Some particular terms in the expansion (18)

1. *The term* $(p=0, q=0)$. In this case we just have

$$\langle P_{1 \to 1}^{(0,0)}(t) \rangle = 1. \tag{20}$$

2. *The term* $(p=1, q=0)$. From (13) we have

$$\langle P_{1 \to 1}^{(1,0)}(t, a_i) \rangle = \int_0^t dt_2 \int_0^{t_2} dt_1 \left[e^{i\Delta_0(t_2-t_1)} \sum_{b_1} \langle V_{a_i b_1}^2 \rangle e^{i(\epsilon_{b_1} - \epsilon_{a_i})(t_2-t_1)} \right]. \tag{21}$$

We now concentrate on the calculation of the quantity in square brackets which, as we shall see, appears systematically in the more complicated terms. The sum over b_1 can be replaced by an integral if the exponential varies only negligibly from one level ϵ_{b_1} to the next; for this to happen, one needs times such that

$$Dt \ll 1, \tag{22a}$$

where D is the mean level spacing. If $1/D$ is interpreted as the Poincarè recurrence time

t_p, we thus need

$$t \ll t_p, \qquad (22b)$$

which is certainly satisfied. We shall neglect the fact that for the single term $a_i = b_1$, $\langle V_{a_ib_1}^2 \rangle$ is twice as big as for an off-diagonal term [see equation (14)]; it can be easily checked that the relative error that we make is of the order of $t/\rho(\epsilon_i) \sim t/t_p$, which is negligible. We thus use (17) and the comment made right after that equation to write ($\tau \equiv t_2 - t_1$)

$$\sum_{b_1} \langle V_{a_ib_1}^2 \rangle e^{i(\epsilon_{a_i} - \epsilon_{b_1})\tau}$$
$$= v_{a_i}^2 \rho(\epsilon_{a_i}) \int_0^\infty w_\Delta(\epsilon_{a_i} - \epsilon_{b_1}) e^{-i(\epsilon_{a_i} - \epsilon_{b_1})\tau} d\epsilon_{b_1} \qquad (23)$$
$$= \frac{\Gamma_{a_i}}{4\pi} \int w_\Delta(x) e^{ix\tau} dx \equiv \frac{\Gamma_{a_i}}{4\pi} \tilde{w}(\tau)$$

We shall consider initial bath states a_i such that the full energy interval $\epsilon_{a_i} \pm \Delta$, to which they are connected by the interaction, is not cut out by the lower bound of the spectrum; for a given temperature, we shall assume that the relevant ϵ_{a_i}'s fulfill this condition, our analysis would thus be valid above a certain minimum temperature T_0 (depending on the specific structure of the matrix V), in order to avoid any "threshold effect." We shall see that above T_0, at variance with the degenerate case of reference 8, the result does depend on Δ. Under these conditions, $\tilde{w}(\tau)$ of equation (23) is real, symmetric in τ, appreciable only inside the interval $\Delta\tau \sim 1/\Delta$ and has the property

$$\int_{-\infty}^\infty \tilde{w}(\tau) d\tau = \pi w(0). \qquad (24)$$

We recall that $\tilde{w}(\tau)$ has to be used inside the time integral of equation (21). Therefore, for times much larger than $1/\Delta$ [which was interpreted in equation (15) as the duration of a collision], $\tilde{w}(\tau)$ behaves like a δ-function and we can write

$$e^{\mp i\Delta_0\tau} \sum_{b_1} \langle V_{a_ib_1}^2 \rangle e^{i(\epsilon_{a_i} - \epsilon_a)\tau} = \frac{\Delta}{\Delta \pm i\Delta_0} \Gamma(\epsilon_{a_i}) \delta(\tau). \qquad (25)$$

To summarize, the times involved in the problem are assumed to fulfill the inequalities

$$t_{\text{coll}} \ll t \ll t_p. \qquad (26)$$

We thus get, for the $(1,0)$ term of (21), the expression

$$\langle P_{1\to 1}^{(1,0)}(t,a_i) \rangle = -\frac{\Gamma_{a_i}}{2} \frac{t\Delta}{2(\Delta - i\Delta_0)} \qquad (27)$$

the last factor $1/2$ arising from the fact that the t_1 integration in (21) goes only up to t_2, thus covering "half of the δ-function".

In preparation for the analysis of the more complicated terms, it will be useful to introduce a graphical representation for the term we have just calculated. In Figure 2 and the following ones, we have indicated two time intervals, from 0 to t, which should

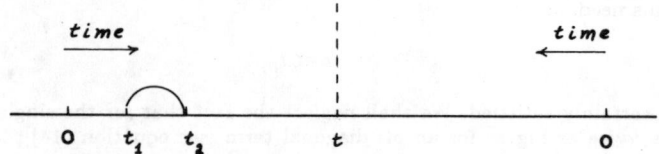

FIGURE 2. Graphical representation of the $(p=1, q=0) = (1,0)$ term of equation (20).

contain, respectively, the $2p$ and $2q$ ordered time variables of equation (13). In the present case, equation (21) just contains t_1 and t_2, which have been indicated in Figure 2. The line joining them, to be called a *contraction*, indicates the ensemble average of the two corresponding V's [remember that in equation (11) every t_i has a $H_I(t_i)$ associated with it]. It is clear that our result (27) can be written at once from the diagram, by following the rules:

a) Assign to the contraction a factor $\Gamma_{a_i}/2$.
b) Assign to the contraction an extra factor $1/2$, whose origin is explained right after equation (27).
c) The two times t_1 and t_2 are reduced to a single one [due to the δ-function (25)], to be integrated from 0 to t; that integration gives a factor $t\Delta/(\Delta - i\Delta_0)$.

Finally, we see from equation (19) that the $(p=0, q=1)$ term gives a contribution which is the complex conjugate of (27), so that, to *first order we have*

$$\langle P^{a_i}_{1 \to 1}(t) \rangle = 1 - \frac{\Gamma_{a_i}}{2} \frac{\Delta^2 t}{\Delta^2 + \Delta_0^2} + \cdots \tag{28}$$

and

$$\langle P^{a_i}_{1 \to -1}(t) \rangle = \frac{\Gamma_{a_i}}{2} \frac{\Delta^2 t}{\Delta^2 + \Delta_0^2} + \cdots \tag{29}$$

Result (29) coincides with that obtained from the "golden rule" of Quantum Mechanics,[15] where a restriction on time similar to (26) also appears, the role of Δ being played by the energy interval over which $V^2_{a;a}$ varies appreciably. The structure of the matrix V which is not required to obtain the golden rule, turns out to be very useful in the evaluation of the higher-order terms of the series, as we shall see.

3. *The higher-order terms.* These terms will be calculated using a well known theorem of statistics:[7-9] in order to calculate the average of a product of zero-centered Gaussian variables one contracts those variables in pairs and sums over all possible pair-contraction patterns.

As an example, the $(1,1)$ term involves the three diagrams shown in Figure 3. We shall need below the relation

$$\langle V_{ab} V_{bc} \rangle = \langle V^2_{ab} \rangle \delta_{ac}, \tag{30}$$

which is a consequence of the basic assumption (14).

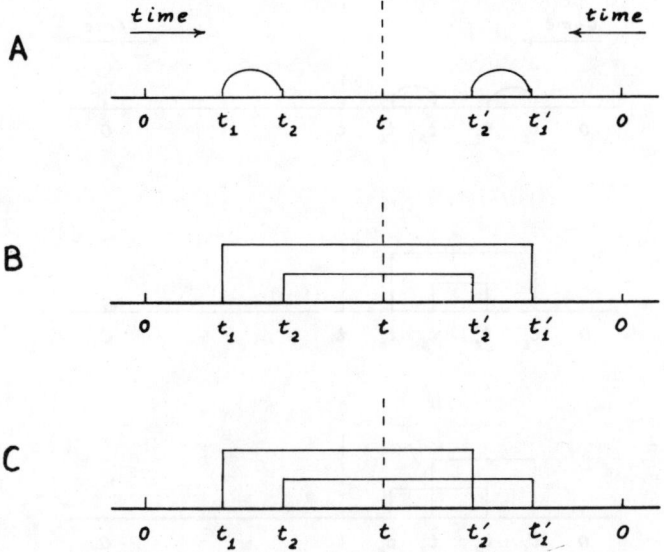

FIGURE 3. The three diagrams arising from the $(p = 1, q = 1) = (1, 1)$ term in the expansion of equations (12) and (18).

Using (30) we can write diagram A as

$$\langle P_{1\to1}^{(1,1)}(t, a_i)\rangle_A = \int_0^t dt_2 \int_0^{t_2} dt_1 \int_0^t dt'_2 \int_0^{t'_2} dt'_1$$
$$\left[e^{i\Delta_0(t_2-t_1)} \sum_{b_1} \langle V_{a_ib_1}^2 \rangle e^{i(\epsilon_{b_1}-\epsilon_{a_i})(t_2-t_1)} \right] \quad (31)$$
$$\times \left[e^{-i\Delta_0(t'_2-t'_1)} \sum_{b'_1} \langle V_{a_ib'_1}^2 \rangle e^{-i(\epsilon_{b'_1}-\epsilon_{a_i})(t'_2-t'_1)} \right],$$

which shows that the basic block of equation (25) makes its appearance again. We thus apply to each contraction the rules found above to write

$$\langle P_{1\to1}^{(1,1)}(t, a_i)\rangle_A = \left(\frac{\Gamma_{a_i}}{2}\right)^2 \frac{\Delta^2 t^2}{2^2} \left(\frac{1}{\Delta + i\Delta_0}\right) \left(\frac{1}{\Delta - i\Delta_0}\right). \quad (32)$$

Diagram B involves *cross contractions*, i.e. contractions that cross the dashed line. Using (30), one can see that the block (25) occurs again. This will be a common feature of all the remaining terms! Rules (a) and (b) above are valid for any contraction. The rule (c) is now the following: due to the δ-functions of (24), $t'_1 = t_1$ and $t'_2 = t_2$ so that we have two ordered times, t_1 and t_2, to be integrated from 0 to t, giving $(t^2/2)[(\Delta/(\Delta +$

54

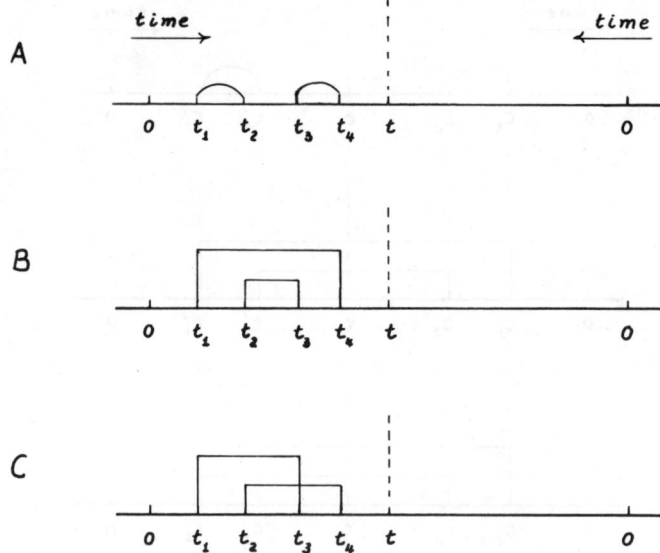

FIGURE 4. The three diagrams arising from the (4, 0) term in the expansions (13) and (18).

$i\Delta_0)) + (\Delta/(\Delta - i\Delta_0))]^2$. The result is thus

$$\langle P^{(1,1)}_{1\to 1}(t, a_i)\rangle_B = \left(\frac{\Gamma_{a_i}}{2}\right)^2 \frac{\Delta^2 t^2}{2^2 \cdot 2}\left(\frac{2\Delta}{\Delta^2 + \Delta_0^2}\right)^2. \quad (33)$$

This shows that the rule (c) assigns a factor $(\Delta/(\Delta - i\Delta_0))$ or $(\Delta/(\Delta + i\Delta_0))$ for each non-cross contraction depending if it contracts t_i with t_{i+1} or t'_i with t'_{i+1} and a factor $[(\Delta/(\Delta + i\Delta_0)) + (\Delta/(\Delta - i\Delta_0))]$ for each cross-contraction.

Diagram C involves the special feature of *crossing lines which*, due to the basic assumption (14), kill one summation, in comparison with diagrams A and B; its contribution can be easily seen to be of order $t/\rho(\epsilon_{a_i}) \sim t/t_p$, relative to A or B, and hence negligible. The fact that *contractions with crossing lines are negligible* is a well known rule.[7-9]

4. *The term* $(2, 0)$. This term gives rise to the three diagrams of Figure 4. For diagram A we apply rules (a) and (b) above and then realize that we are left with two ordered times to be integrated from 0 to t giving $(t^2/2!)(\Delta/(\Delta - i\Delta_0))^2$. The result is then

$$\langle P^{(2,0)}_{1\to 1}(t, a_i)\rangle_A = \left(\frac{\Gamma_{a_i}}{2}\right)^2 \frac{\Delta^2 t^2}{2^2 \cdot 2!}\left(\frac{1}{\Delta - i\Delta_0}\right)^2. \quad (34)$$

Diagram B does not contribute, because the time ordering, in addition with the δ-functions that occur, annihilates the integration domain. We thus find the rule that *non-cross contractions can only be contiguous*.

Finally, diagram C is negligible because it involves intersecting lines.

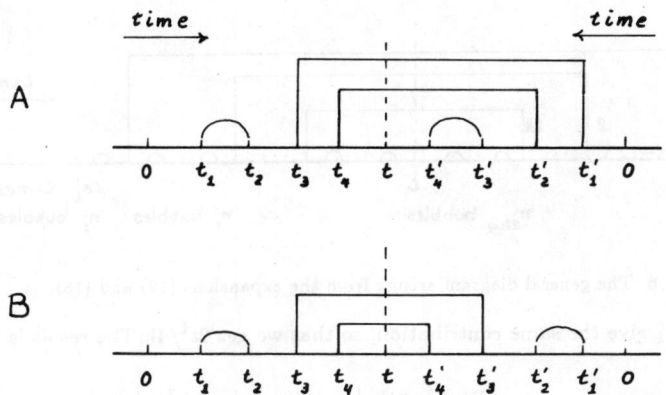

FIGURE 5. Two of the diagrams arising from the (2, 2) term in the expansions (12) and (18).

5. The term $(p, 0)$. From the above considerations, the only diagram that contributes to the term $(p, 0)$ contains p contiguous non-cross contractions, or "bubbles", thus giving

$$\langle P_{1\to 1}^{(p,0)}(t, a_i)\rangle = (-)^p \left(\frac{\Gamma_{a_i}}{2}\right)^p \frac{t^p}{2^p p!} \left(\frac{\Delta}{\Delta - i\Delta_0}\right)^2. \qquad (35)$$

6. The term $(p = 2, q = 2)$. We consider the two diagrams of Figure 5, taken from those that contribute to this term.

Diagram A. From rules (a) and (b) above, we have the following contributions:

$$4 \text{ contractions} \implies \left(\frac{\Gamma_{a_i}}{2}\right)^4 \frac{1}{2^4}.$$

The equivalent of rule (c) is now the following: due to the cross contractions we get δ-functions that can be used to eliminate t_1' and t_2', setting them equal to t_3 and t_4, respectively, and bringing them to the LHS of the diagram; we also have $t_3' = t_4'$, which can again be brought to the LHS of the diagram and integrated from t_4 to t. We are left with 4 ordered times on the LHS, to be integrated from 0 to t giving

$$\frac{t^4}{4!} \left|\frac{\Delta}{\Delta - i\Delta_0}\right|^2 \left(\frac{\Delta}{\Delta + i\Delta_0} + \frac{\Delta}{\Delta - i\Delta_0}\right)^2.$$

We finally have

$$\langle P_{1\to 1}^{(2,2)}(t, a_i)\rangle_A = \left(\frac{\Gamma_{a_i}}{2}\right)^4 \frac{t^4}{2^4 \cdot 4!} \left|\frac{\Delta}{\Delta - i\Delta_0}\right|^2 \left(\frac{2\Delta^2}{\Delta^2 + \Delta_0^2}\right)^2. \qquad (36)$$

Diagram B. We again use the δ-functions to bring all the times to the LHS. We have two times $t_1 = t_2$ and $t_2' = t_1'$ varying independently (*i.e.* they are not ordered) from 0 to t_3, and then the two ordered times t_3 and t_4, to be integrated from 0 to t. If t_1 and t_1' were ordered, the whole time integral would give $t^4/4!$; however, the two possibilities

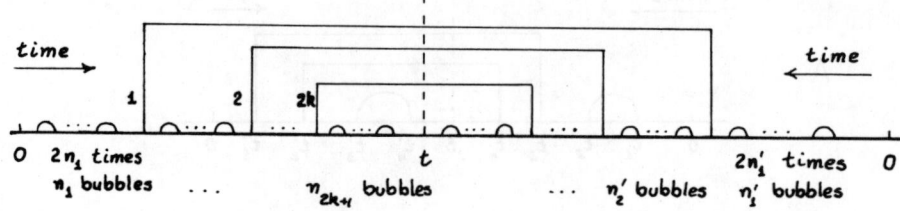

FIGURE 6. The general diagram arising from the expansions (12) and (18).

$t_1 < t'_1$ and $t_1 > t'_1$ give the same contribution, so that we get $2t^4/4!$. The result is then

$$\langle P^{(2,2)}_{1\to 1}(t,a_i)\rangle_A = \left(\frac{\Gamma_{a_i}}{2}\right)^4 \frac{2t^4}{2^4 \cdot 4!}\left|\frac{\Delta}{\Delta - i\Delta_0}\right|^2 \left(\frac{2\Delta^2}{\Delta^2 + \Delta_0^2}\right)^2. \qquad (37)$$

3.2. The general rules. Evaluation of the survival probability

We now collect the rules that we have found from the above analysis. First, the qualitative rules:

1) Only contractions with non-intersecting lines contribute.
2) Non-cross contractions can only be contiguous (bubbles).
3) The number of cross contractions must be even.

The structure of the general diagram is illustrated in Figure 6. It has the following characteristics:

a) There are $2p$ times on the LHS and $2q$ on the RHS.
b) There are $(2p + 2q)/2 = p + q$ contractions. Out of these, $2k$ are cross contractions; there are left $(2p - 2k)/2 = p - k$ bubbles on the LHS and $(2q - 2k)/2 = q - k$ on the RHS.
c) We have the relations

$$\sum_{i=1}^{2k+1} n_i + k = p, \qquad \sum_{i=1}^{2k+1} n'_i + k = q. \qquad (38\text{a, b})$$

We now have the following numerical contributions to the (p,q) term:

1) A factor $(-)^{p+q}$
2) Each contraction gives a factor

$$\frac{1}{2}\frac{\Gamma_{a_i}}{2} \implies \frac{1}{2^{p+q}}\left(\frac{\Gamma_{a_i}}{2}\right)^{p+q}.$$

3) Time integrations. Each bubble in the LHS gives a factor $\Delta/(\Delta - i\Delta_0)$, each bubble

in the RHS gives a factor $\Delta/(\Delta+i\Delta_0)$ and each cross contraction a factor

$$\left(\frac{\Delta}{\Delta-i\Delta_0}+\frac{\Delta}{\Delta+i\Delta_0}\right)\Longrightarrow\left(\frac{\Delta}{\Delta+i\Delta_0}\right)^{q-k}\left(\frac{\Delta}{\Delta-i\Delta_0}\right)^{p-k}\left(\frac{2\Delta^2}{\Delta^2+\Delta_0^2}\right)^{2k}$$

We bring all the times to the LHS. As a result, we have n_1+n_1' bubbles between 0 and the first cross contractions, n_2+n_2' between the first and the second cross contraction, etc. The number of time variables to be integrated from 0 to t is thus

$$(n_1+n_1')+(n_2+n_2')+\cdots+(n_{2k+1}+n_{2k+1}')+2k=p+q. \tag{39}$$

If these $p+q$ time variables were *all* ordered, the final time integral from 0 to t would give $t^{p+q}/(p+q)!$. This is not the case, though, because: the $2k$ times associated with the cross contractions are indeed ordered; between the $(i-1)$-th and the i-th cross contraction we have n_i+n_i' bubbles, of which n_i are ordered among themselves, n_i' are also ordered among themselves, but the n_i ones can be in any position relative to the n_i' ones: the total number of possibilities for the i-th interval is thus the number of permutations of n_i+n_i' objects, disregarding the permutations of n_i and n_i' object separately; i.e., $(n_i+n_i')!/n_i!n_i'!$.

The final result of the time integrations is thus

$$\frac{t^{p+q}}{(p+q)!}\prod_{i=1}^{2k+1}\binom{n_i+n_i'}{n_i}\left(\frac{\Delta}{\Delta-i\Delta_0}\right)^{p-k}\left(\frac{\Delta}{\Delta+i\Delta_0}\right)^{q-k}\left(\frac{2\Delta^2}{\Delta^2+\Delta_0^2}\right)^{2k}$$

Collecting the above results and summing over all allowed diagrams, we find the survival probability of equation (18), before averaging over initial states, as

$$\langle P_{1\to 1}^{a_i}(t)\rangle=\sum_{p,q,k}(-)^{p+q}\frac{(\Gamma_{a_i}/2)^{p+q}}{2^{p+q}(p+q)!}\left(\frac{\Delta t}{\Delta-i\Delta_0}\right)^{p-k}\left(\frac{\Delta t}{\Delta+i\Delta_0}\right)^{q-k}\left(\frac{2t\Delta^2}{\Delta^2+\Delta_0^2}\right)^{2k}$$
$$\times\sum_{\{n_i\}\{n_i'\}}{}'\binom{n_i+n_1'}{n_1}\binom{n_2+n_2'}{n_2}\cdots\binom{n_{2k+1}+n_{2k+1}'}{n_{2k+1}}.$$
(40)

The prime in the last summation indicates that the restrictions (38) have to be enforced.

Using equation (19) of reference 16 we can evaluate the last sum in equation (40), with the result

$$\sum_{\{n_i\}\{n_i'\}}{}'=\binom{p+q}{2k}\binom{p+q-2k}{p-k}. \tag{41}$$

The remaining sums in (40) can also be performed, the result being

$$\langle P_{1\to 1}^{a_i}(t)\rangle=\frac{1}{2}\left[1+\exp\left(-\frac{\Gamma_{a_i}\Delta^2}{\Delta^2+\Delta_0^2}t\right)\right]. \tag{42}$$

Averaging over initial states as in (18) we finally have

$$\langle P_{1\to 1}(t)\rangle = \frac{1}{2}\left[1 + \left\langle \exp\left(-\frac{\Gamma(\epsilon)\Delta^2}{\Delta^2+\Delta_0^2}t\right)\right\rangle_\beta\right], \qquad (43)$$

where $\langle\cdots\rangle_\beta$ indicates the thermal average

$$\left\langle \exp\left(-\frac{\Gamma(\epsilon)\Delta^2}{\Delta^2+\Delta_0^2}t\right)\right\rangle = \frac{1}{Z}\int_0^\infty \exp\left(-\frac{\Gamma(\epsilon)\Delta^2}{\Delta^2+\Delta_0^2}\right)\exp(-\beta\epsilon)\rho(\epsilon)\,d\epsilon, \qquad (44)$$

Z being the bath partition function of equation (8).

Equation (43) is our main result. A discussion of some of its properties is given in the next section.

4. PROPERTIES OF THE SURVIVAL PROBABILITY

We first observe that, if we expand the exponential in (42), we get back the first order result (28) provided by the golden rule.

On the other hand, as $t \to \infty$, the survival probability tends to $1/2$, indicating that both states of the system, $\alpha = \pm 1$, become equally populated.

From (43) we can calculate the *transition probability* as

$$\langle P_{1\to -1}(t)\rangle = \frac{1}{2}\left[1 - \left\langle \exp\left(-\frac{\Gamma(\epsilon)\Delta^2}{\Delta^2+\Delta_0^2}t\right)\right\rangle_\beta\right]. \qquad (45)$$

We can also calculate the *polarization* $\pi(t)$, defined as

$$\pi(t) = \langle\sigma_z\rangle, \qquad (46)$$

where the bracket denotes a quantum-mechanical plus an ensemble average. Writing $\sigma_z = |1\rangle\langle 1| - |-1\rangle\langle -1|$, we can express the polarization as

$$\pi(t) = \langle P_{1\to 1}(t)\rangle - \langle P_{1\to -1}(t)\rangle. \qquad (47)$$

Using (43) and (45) we then find

$$\pi(t) = \left\langle \exp\left(-\frac{\Gamma(\epsilon)\Delta^2}{\Delta^2+\Delta_0^2}t\right)\right\rangle_\beta. \qquad (48)$$

We now center our discussion on this last quantity.

It is clear form (42) that before averaging over initial states we get an exponential decay for the polarization, for any a_i. However, if Γ_{a_i} depends on ϵ_i, each initial state contributes with its own decay probability to the average (48) and the result is, in general, a non-exponential decay law.

Therefore, if $\Gamma(\epsilon)$ is a constant, independent of ϵ, i.e.

$$\Gamma(\epsilon) = \Gamma, \qquad (49)$$

the polarization (48) shows the exponential decay

$$\pi(t) = \exp-\left(\frac{\Gamma\Delta^2}{\Delta^2+\Delta_0^2}t\right). \qquad (50)$$

For a non-constant $\Gamma(\epsilon)$, even a slow energy dependence may be important in distorting the exponential (50), since $\Gamma(\epsilon)$ occurs in the exponent in equation (48). More explicit results can be obtained with more specific form for $\Gamma(\epsilon)$ and $\rho(\epsilon)$.

5. SUMMARY AND CONCLUSIONS

We were able to extend the random-matrix model used in reference 8 to the case were the total Hamiltonian contains a term which describes the two-level-system dynamics. As in reference 8, we considered a "local GOE" for the system-bath interaction and calculated the ensemble average of the survival probability for times larger than the duration of a collision and smaller than the recurrence or Poincarè time. It was also possible to sum the whole series into a compact functional form. The final result is expressed as an average over the initial states whose level density depends on the particular thermal bath of the problem. As we mention before, the survival probability depends both on the width of the weight factor that appears in the definition of the "local GOE" in equation (14) and on the parameter Δ_0 related with the strength of the magnetic field around which the spin-1/2 particle is assumed to precess. At the end, the total effect of the additional term in the Hamiltonian is to multiply t by the factor $\Delta^2/(\Delta^2+\Delta_0^2)$. In the particular case of $\Gamma(\epsilon)$ constant, the survival probability becomes a simple function [see equation (50)] and clearly the effect of the factor $\Delta^2/(\Delta^2+\Delta_0^2)$ is to reduce, for a given time t, the probability of a spin flip process. There is a number of applications of our result which are under investigation.

REFERENCES

1. *Electron-Spin Relaxation in Liquids*, edited by L.T. Muus and P.W. Atkins (Plenum Press, New York, 1972).
2. Harris, R.A. and Silbey, R., *J. Chem. Phys.* **78**, 7330 (1983); Nitzan, A. and Silbey, R.J., *J. Chem. Phys.* **60**, 4070 (1974).
3. Rivier, N. and Coe, T.J., *J. Phys. C.* **10**, 4471 (1977).
4. Chakravarty, S. and Leggett, A.J., *Phys. Rev. Lett.* **52**, 5 (1984); Leggett, A.J. et al., *Rev. Mod. Phys.* **59**, 1 (1987) (and references therein).
5. Pfeifer, P., *Symmetries and Properties of Nonrigid Molecules: A comprehensive Survey*, p. 379, edited by J. Maruani and J. Serre (Elsevier, Amsterdam).
6. Cohen-Tanaudji, C., *Frontiers in Laser Spectroscopy*, Vol. 1, p. 3, edited by R. Balian, S. Haroche and S. Libermân (North-Holland).
7. Brody, T.A., Flores, J., French, J.B., Mello, P.A., Pandey, A. and Wong, S.S.M., *Rev. Mod. Phys.* **53**, 385 (1981).
8. Mello, P.A., Pereyra, P. and Kumar, N., *J. Stat. Phys.* **51**, (1988).
9. Agassi, D., Weidenmüller, H.A. and Mantzouranis, G., *Phys. Rep.* **22**, 145 (1975); Agassi, D., Ko, C.M. and Weidenmüller, H.A., *Ann. Phys. (NY)* **107**, 140 (1977); Weidenmüller, H.A., *Theoretical Methods in Medium-Energy and Heavy-Ion Physics*, edited by K.W. McVoy and W.A. Friedman (Plenum Press, New York, 1978).
10. Anderson, P.W., *J. Phys. Soc. Japan* **9**, 316 (1954).
11. Kubo, R., *J. Phys. Soc. Japan* **17**, 1100 (1962).
12. Fox, R.F., *Phys. Rep.* **48**, 179 (1978), part II.

13. van Kampen, N.G., *Physica* **74**, 215 (1974); *Physica* **74**, 239 (1974).
14. Terwiel, R.H., *Physica* **74**, 248 (1974).
15. Messiah, A., *Quantum Mechanics, Vol. II*, p. 736, G.M. Temmer tr. (John Wiley and Sons, Inc., 1966).
16. Netto, E., *Lehrbuch der Kombinatorik, Ch. 13*, (Chelsea Pub. Co., New York, 1927).

MACROSCOPIC APPROACH TO DISORDERED CONDUCTORS

P.A. Mello*

Instituto de Física, Universidad Nacional Autónoma de México,
apartado postal 20-364, 01000 México, D.F., MEXICO

In the usual *microscopic* approaches to disordered conductors at zero temperature, a statistical law for the individual scatterers ("microscopic" quantities) is assumed, and then used to evaluate the various averages of interest for the full conductor ("macroscopic" quantities). Since many of the results so obtained do not depend on the details of the distribution of the microscopic quantities, we shall adopt the philosophy of directly studying statistical distributions associated with the *full* conductor: the resulting theory will thereby be named *macroscopic*.[1-3]

In the scattering approach, the disordered system (a piece of wire) is sandwiched between two perfect leads, where the scattering states, at the Fermi energy, define the N channels; each channel can carry two waves, propagating in opposite directions. The wave function outside the scattering system is thus specified by a $2N$-component vector, whose first N components are the amplitudes of the waves travelling to the right, and the remaining components are the N amplitudes travelling to the left. By definition, the $2N \times 2N$ *transfer matrix* R relates the vector on the right with that on the left of the system. Flux conservation and time-reversal invariance require $R \sum_z R^+ = \sum_z$, $R^* = \sum_x R \sum_x$, where

$$\sum_z = \begin{pmatrix} 1 & 0 \\ 0 & -1 \end{pmatrix}, \quad \sum_x = \begin{pmatrix} 0 & 1 \\ 1 & 0 \end{pmatrix}$$

have the structure of Pauli matrices, 1 indicating the $N \times N$ unit matrix. With these restrictions, an R matrix has $N(2N+1)$ independent parameters and can always be represented in the form[3]

$$R = \begin{pmatrix} u & 0 \\ 0 & u^* \end{pmatrix} \begin{pmatrix} \sqrt{1+\lambda} & \sqrt{\lambda} \\ \sqrt{\lambda} & \sqrt{1+\lambda} \end{pmatrix} \begin{pmatrix} v & 0 \\ 0 & v^* \end{pmatrix}, \qquad (1)$$

where u, v are arbitrary $N \times N$ unitary matrices and λ is a real, diagonal matrix with N arbitrary positive elements $\lambda_1, \ldots, \lambda_N$. As an application of the above parametrization, we mention that the $N \times N$ transmission and reflection matrices are given by

$$t = u \frac{1}{\sqrt{1+\lambda}} v \qquad (2)$$

*Also at Departamento de Física, UAM-Iztapalapa, and fellow of the Sistema Nacional de Investigadores.

$$r = -v^\mathsf{T}\sqrt{\frac{\lambda}{1+\lambda}}v, \tag{3}$$

in terms of which the transmission and reflection coefficients are $T_{ab} = |t_{ab}|^2$ and $R_{ab} = |r_{ab}|^2$. When the channels are fed with N incoherent unit fluxes, the total transmission coefficient $T = \sum_a T_a$ into all channels is given by $T = \sum_a (1+\lambda_a)^{-1}$. This is a very important quantity, since in the metallic regime, where the sample dimensions are much greater than the mean free path ℓ and each $T_a \ll 1$, the conductance g (including spin) is given by[4-6] $g = 2T$.

A collection or ensemble of random conductors of length L is described by an ensemble of R matrices, whose differential probability $dP_L(R) = p_L(R)\, d\mu(R)$ depends parametrically upon L. Here $d\mu(R)$, the *invariant measure* associated with the group of R's is given by[3,7] $J(\lambda)\prod_a d\lambda_a\, d\mu(u)d\mu(v))$, where $J(\lambda) = \prod_{a<b}|\lambda_a - \lambda_b|$ and $d\mu(u)$ (and $d\mu(v)$) is the invariant measure of the unitary group $U(N)$.

The probability density $p_L(R)$ must satisfy an *important combination law*. If we put together two wires of lengths L and δL, with probability densities p_L and $p_{\delta L}$, the resulting probability density is given by the convolution $p_{L+\delta L} = p_L * p_{\delta L}$. If we knew the "building block" $p_{\delta L}(R')$, we would construct $p_L(R)$ for arbitrary lengths by successive convolutions.

In a complete theory, we should be able to impose the appropriate physical requirements that would allow the unique determination of the building block. At the present stage, we propose an "ansatz" for $p_{\delta L}(R')$. We choose the statistical distribution that maximizes Shannon's *information entropy*[8]

$$\mathcal{L}[p_{\delta L}] = -\int p_{\delta L}(R')\ln p_{\delta L}(R')\,d\mu(R'), \tag{4}$$

constrained by the conditions that $p_{\delta L}$ be normalized and that the average

$$\frac{N^{-1}\langle\mathrm{tr}\,\lambda'\rangle_{\delta L}}{\delta L} \equiv \frac{1}{\ell} \tag{5}$$

be fixed. For small λ', (5) represents the reflection probability per unit length, which is the inverse mean free path $1/\ell$ for backward scattering. The resulting distribution can be written as

$$p_{\delta L}(R') = e^{\mu - \nu\,\mathrm{tr}\,\lambda'}, \tag{6}$$

where μ, ν are Lagrange multipliers.

Notice that (6) is *isotropic*, i.e., independent of the unitary matrices u, v of (1). One can prove[3] that the convolution of two isotropic functions is again isotropic, so that the resulting $p_L(R)$ is only a function of $\lambda = \lambda_1,\ldots,\lambda_N$.

The ansatz (6), introduced in the *combination requirement*, gives, for the joint probability density $w_L(\lambda) \equiv p_L(\lambda)J(\lambda)$ of $\lambda_1\cdots\lambda_N$, the Fokker-Planck or diffusion equation[3] (with $s \equiv L/\ell$)

$$\frac{\partial w_s(\lambda)}{\partial s} = \frac{2}{N+1}\sum_{a=1}^{N}\frac{\partial}{\partial \lambda_a}\left[\lambda_a(1+\lambda_a)J(\lambda)\frac{\partial}{\partial \lambda_a}\frac{w_s(\lambda)}{J(\lambda)}\right]. \tag{7}$$

From equation (7) one could calculate, in principle, the "evolution" with length of the average of any quantity of interest. Before going into that, though, we wish to remark that the isotropy assumption mentioned above has very important consequences. Without making use of any specific statistical distribution $p_L(\lambda)$, *the isotropy assumption* alone gives the structure of the averages and covariances of transmission and reflection factors, as a function of channel indices.[9] It is the specific value of the various coefficients which depends on the solution of the diffusion equation (7).

The average of the transmission coefficient $T_{ab} = |t_{ab}|^2$, t_{ab} being the ab matrix element of equation (2), is given by

$$\langle T_{ab}\rangle_s = \sum_{\alpha\alpha'} M^{a\alpha}_{a\alpha'} M^{\alpha b}_{\alpha' b} \langle \sqrt{\tau_\alpha \tau_\beta}\rangle_s, \tag{8}$$

where $\tau_\alpha = (1+\lambda_\alpha)^{-1}$. The last factor in equation (8) is an average performed with $w_s(\lambda)$, for which we need not be more specific for the time being. The factors M occurring in equation (8) are a particular case of the general average[10, 11]

$$M^{a_1\alpha_1,\ldots,a_m\alpha_m}_{a'_1\alpha'_1,\ldots,a'_m\alpha'_m} = \langle (u_{a'_1\alpha'_1}\cdots u_{a'_m\alpha'_m})(u_{a_1\alpha_1}\cdots u_{a_m\alpha_m})^*\rangle_0, \tag{9}$$

performed with the invariant measure of the unitary group (indicated by the index 0).

In reference 10 it is shown that

$$M^{a\alpha}_{a'\alpha'} = \frac{\delta_{a'a}\delta_{\alpha'\alpha}}{N}, \tag{10}$$

so that equation (8) becomes

$$\langle T_{ab}\rangle_s = N^{-2}\langle T\rangle_s, \tag{11}$$

where $T = \sum_{ab} T_{ab}$ is the total transmission factor into all channels, when the incident channels are fed with N incoherent unit fluxes.

Next we calculate, from equation (2), the crossed second moment

$$\langle T_{ab} T_{a'b'}\rangle_s = \sum_{\alpha\beta\alpha'\beta'} M^{a\alpha,a'\beta}_{a\alpha',a'\beta'} M^{\alpha b,\beta b'}_{\alpha' b,\beta' b'} \langle (\tau_\alpha \tau_\beta \tau_{\alpha'} \tau_{\beta'})^{1/2}\rangle_0. \tag{12}$$

In references 10 and 11 the M-coefficients of equation (12) are calculated. In terms of them, the covariance $C^T_{ab,a'b'} = \langle T_{ab}T_{a'b'}\rangle - \langle T_{ab}\rangle\langle T_{a'b'}\rangle$ can be expressed as

$$\begin{aligned}C^T_{ab,a'b'} = \frac{1}{(N^2-1)^2} &\left\{ \left[\left(1+\frac{1}{N^2}\right)\langle T^2\rangle_s - \frac{2}{N}\langle T_2\rangle_s\right]\delta_{aa'}\delta_{bb'}\right.\\ &+ \left[\left(1+\frac{1}{N^2}\right)\langle T_2\rangle_s - \frac{2}{N}\langle T^2\rangle_s\right](\delta_{aa'}+\delta_{bb'})\\ &\left. + \left[\operatorname{var} T + \frac{1}{N^2}\langle T^2\rangle_s + \frac{2}{N^2}\langle T\rangle_s^2 - \frac{1}{N^4}\langle T\rangle_s^2 - \frac{2}{N}\langle T_2\rangle_s\right]\right\}.\end{aligned} \tag{13}$$

Here we have defined $T_2 = \sum_a (1+\lambda_a)^{-2}$. Equation (13) is exact. As a check, we can easily verify that the sum of (13) over a, a', b, b' gives precisely $\operatorname{var} T$.

In reference 12, equations (3), three types of terms are also obtained: setting $W \ll L$

(**quasi-1D** systems), they are seen to have essentially the structure provided by the δ-functions of our equation (13). The difference is that our Kronecker deltas (that we can write as $\delta_{aa'} = \delta_{\Delta q_a = 0}$ with $\Delta q_a = |q_a - q_{a'}|$, q_a being the transverse wave-vector labeling the channel (the eigenmode) a) are replaced by some "smeared" (on a distance $\Delta q \sim 1/L$) δ-functions in reference 12.

In a similar way one finds, for the average of the reflection coefficient R_{ab}

$$\langle R_{ab} \rangle_s = (1 + \delta_{ab}) \frac{\langle R \rangle_s}{N(N+1)}, \tag{14}$$

R being the total reflection coefficient $\sum_{ab} R_{ab}$. Result (14) means that backward scattering to the same channel is enhanced by a factor 2 as compared with the scattering to any other channel. Except for a smeared out cone, this is precisely the *enhanced backscattering* predicted by weak localization theory.[13]

Similarly, one can find the structure in the channel indices of the covariance of the reflection coefficients.[9]

We now go back to the diffusion equation (7), which governs the "evolution" of our probability distribution.

Multiplying both sides of (7) by T^p and integrating, one obtains the evolution equation for the p-th moment of the total transmission coefficient as[3, 14]

$$(N+1)\partial_s \langle T^p \rangle_s = \langle -pT^{p+1} - pT^{p-1}T_2 + 2p(p-1)T^{p-2}(T_2 - T_3) \rangle_s, \tag{15}$$

where $T_k = \sum_k (1 + \lambda_a)^{-k}$.

We notice that on the RHS of (15) there appear quantities other than $\langle T^p \rangle_s$, so that their evolution equations are needed as well. Since we are interested in the $N \gg 1$ limit, we have studied[14] the solution of the coupled equations as a series in decreasing powers of N. The result is

$$\langle T^p \rangle_s = \frac{N^p}{(1+s)^p} - \frac{ps^3}{3(1+s)^{p+2}} N^{p-1} + p \frac{(11p-9)s^8 + \cdots}{90(1+s)^{p+6}} N^{p-2} + \cdots. \tag{16}$$

For $p = 1$ and $s = L/\ell \gg 1$, we have

$$\langle T \rangle_s = \frac{N\ell}{L} + \cdots, \tag{17}$$

which is Ohm's law, when $N \sim (k_F w)^{d-1}$.

From equation (16) we can also calculate var T, with the result, for $s \gg 1$:

$$\text{var} = \frac{2}{15} + \cdots, \tag{18}$$

which is *independent* of the number N of channels (determined by the cross section of the wire), the length L of the conductor and the mean free path ℓ. The rms of the conductance $g = 2T$ is thus a *universal* number: *i.e.*

$$\text{rms } g = \sqrt{8/15} = 0.730 \cdots. \tag{19}$$

This is the statement of the *universal conductance fluctuations*.[7, 15-18] Result (19) is precisely the value found in reference 16, using microscopic Green's function techniques.

The results of reference 14 can be used to find the coefficients appearing in equation (13) for the covariance of the transmission coefficients. For $N \gg 1$, $s \gg 1$, one finds

$$C^T_{ab,a'b'} = \langle T_{ab}\rangle_s \langle T_{a'b'}\rangle_s \left[\delta_{aa'}\delta_{bb'} + \frac{2}{3\langle T\rangle_s}(\delta_{aa'} + \delta_{bb'}) + \frac{2}{15\langle T\rangle_s^2}\right], \qquad (20)$$

in agreement with equations (3) of reference 12 when $W \ll L$.

To summarize, we have presented a theory of disordered conductors based on the general properties of the scattering system: *flux-conservation*, time *reversal invariance* (in the absence of a magnetic field) and the appropriate *combination law* when two wires are put together. The distribution associated with systems of very small length is selected on the basis of a *maximum-entropy* criterion; the combination law allows then to find the "evolution" of that distribution with the length L: it turns out to be governed by a Fokker-Planck or diffusion equation in N dimensions, where N is the *number of channels*. The results emerging from this approach give a good description of disordered conductors in the *metallic regime* and for *quasi-one-dimensional systems*.

REFERENCES

1. Mello, P.A., *Phys. Rev.* B35, 1082 (1987).
2. Mello, P.A., "Quantum Chaos and Statistical Nuclear Physics" (T.H. Seligman and H. Nishioka, Eds.), Vol. 263, p. 267, Lecture Notes in Physics (Springer-Verlag, Berlin, Heidelberg, 1986).
3. Mello, P.A., Pereyra, P. and Kumar, N., *Ann. Phys.* 181, 290 (1988).
4. Anderson, P.W., *Phys. Rev.* B23, 4828 (1981).
5. Fisher, D.S. and Lee, P.A., *Phys. Rev.* B23, 6851 (1981); Lee, P.A. and Fisher, D.S., *Phys. Rev. Lett.* 47, 882 (1981).
6. Büttiker, M., Imry, Y., Landauer, R. and Pinhas, S., *Phys. Rev.* B31, 6207 (1985).
7. Muttalib, K.A., Pichard, J.L. and Stone, A.D., *Phys. Rev. Lett.* 59, 2475 (1987).
8. Levine, R.D. and Bernstein, R.B., *Modern Theoretical Chemistry*, edited by W.H. Miller (Plenum, New York, 1976), Vol. III; Alhassid, Y. and Levine, R.D., *J. Chem. Phys.* 67, 4321 (1978).
9. Mello, P.A., Akkermans, E. and Shapiro, B., *Phys. Rev. Lett.* (submitted).
10. Gaudin, M. and Mello, P.A., *J. Phys.* G7, 1085 (1981).
11. Mello, P.A., *J. Phys. A* (unpublished).
12. Feng, S., Kane, C., Lee, P.A. and Stone, A.D., *Phys. Rev. Lett.* (to be published).
13. Akkermans, E. and Maynard, R., *J. Phys. Lett* 46, L1045 (1985).
14. Mello, P.A., *Phys. Rev. Lett.* 60, 1089 (1988).
15. Umbach, C.P., Washburn, S., Laibowitz, R.B. and Webb, R.A., *Phys. Rev.* B30, 4048 (1984); Webb, R.A., Washburn, S., Umbach, C.P. and Laibowitz, R.B., *Phys. Rev. Lett.* 54, 2696 (1985); Washburn, S., Umbach, C.P., Laibowitz, R.B. and Webb, R.A., *Phys. Rev.* B32, 4789 (1985).
16. Stone, A.D., *Phys. Rev. Lett.* 54, 2692 (1985); Lee, P.A. and Stone, A.D., *Phys. Rev. Lett.* 55, 1622 (1985); Lee, P.A., Stone, A.D. and Fukuyama, H., *Phys. rev.* B35, 1039 (1987); Feng, S., Kane, C., Lee, P.A. and Stone, A.D. (to be published).
17. Al'tshuler, B.L., *Pis'ma Zh. Eksp. Teor. Fiz.* 41, 530 (1985) [*JETP Lett.* 41, 648 (1985)]; Al'tshuler, B.L. and Khmel'nitskii, D.E., *Pis'ma Zh. Eksp. Teor. Fiz.* 42, 291 (1985) [*JETP Lett.* 42, 359 (1986)]; Al'tshuler, B.L. and Shklovskii, B.I., *Zh. Eksp. Teor. Fiz.* 91, 220 (1986) [*Sov. Phys. JETP* 64, 127 (1986)].
18. Imry, Y., *Europhys. Lett.* 1 249 (1986).

EQUILIBRIUM POLYMERIZATION AS A PHASE TRANSITION

Sandra C. Greer

Department of Chemistry and Biochemistry,
The University of Maryland at College Park, College Park,
MD 20742

ABSTRACT. We review recent experimental studies of equilibrium polymerization in liquid sulfur and its solutions: studies of the static dielectric constant, the shear viscosity, and the binary liquid phase diagram with biphenyl. We compare the experiments to the theoretical predictions of J.C. Wheeler and collaborators, who view equilibrium polymerization as a phase transition of order parameter dimension zero. We find qualitative agreement, but intriguing quantitative discrepancies. We discuss planned work on equilibrium polymerization in living polymers.

1. INTRODUCTION

We consider here the extension of the modern theory of phase transitions, so successful in understanding magnets, fluids, and polymers, to a type of chemical reaction, equilibrium polymerization. We are interested in measuring various properties of such polymerizing systems and comparing these measurements to theoretical predictions. We will show that the theory gives good qualitative predictions, but that there are intriguing discrepancies between experiment and theory.

Equilibrium polymerization means polymerization which takes place in such a way that the ends of the polymer remain active. Monomers are continually being added and removed, and the polymerization process can be reversed by changing the temperature. The classic example of equilibrium polymerization is that in liquid sulfur.[1] Above its melting point at 120°C, sulfur molecules are (mostly) rings of eight atoms, which produce a yellow liquid of low viscosity. If the sulfur is heated further, at 159°C the liquid becomes red and very viscous. This change is attributed to the polymerization of sulfur: the rings open to form diradicals, which react together to form (mostly) very long chains of about 10^5 sulfur atoms. We note, however, that it is possible for the ends of a diradical chain to react together to make a ring. If the polymeric liquid sulfur is cooled below 159°C, it reverts to the yellow monomeric form.

Many organic molecules also undergo equilibrium polymerization.[2] They are called "living polymers" because the polymer ends remain active ("alive"). The mechanisms for these reactions require the presence of an *initiator* which starts the polymerization process. The living polymers are frequently charged species, in which case the mechanisms are referred to as *cationic polymerization* or *anionic polymerization*. Polymerization occurs as the temperature is lowered below a *ceiling temperature*, as opposed to the polymerization of sulfur, which occurs as the temperature is raised. For example, the mechanism for the anionic polymerization of α-methylstyrene as initiated by sodium napththalene is:[3]

1. $M + Na^+, Naph^- = \cdot M^-, Na^+ + Naph$
2. $2(\cdot M^-, Na^+) = (^-M \to M^-) + 2Na^+$
3. $(^-M \to M^-) + M = (^-M \to M \leftarrow M^-)$
4. $(^-M \to M \leftarrow M^-) + M = (^-M \to M \leftarrow M \leftarrow M^-)$

Here M is the monomer, '·' represents a radical, and '→' and '←' represent chemical bonds that are "head-to-head" and "head-to-tail", respectively. Step (1) is the initiation step, forming the activated monomeric radical anion. In step (2), two activated monomers combine in a "head-to-head" pattern to make a dianion. In subsequent steps, the α-methylstyrene monomer adds "head-to-tail" to the dianion to make a polymer. Thus the final poly-α-methylstyrene molecule contains one "mistake" in its pattern, but is otherwise a head-to-tail sequence. Note that since there is a negative charge on both ends of the polymer, the ends of a molecule cannot join to make a ring, and the polymer must consist of chains.

A theory for such polymerizations was constructed by Tobolsky and Eisenberg in 1959.[4] They simply assumed one equilibrium constant for the initiation step (thermal or chemical) and a second for the propagation step (regardless of the polymer size). The temperature dependences of these equilibrium constants led to qualitatively correct predictions of the monomer concentration and average polymer chain length (or degree of polymerization), as functions of temperature. Beginning in 1980, Wheeler and coworkers[5,6,7] noted the analogy between the partition function for the n-vector magnet (where n is the dimension of the order parameter) and the partition function for equilibrium polymerization. They noted that when the polymers are all chains, then the equilibrium polymerization should be described by the $n = 0$ magnet model. The mean field limit of that model was shown to yield the Tobolsky-Eisenberg model. However, it was noted that if the polymers are instead in rings, then the appropriate model is that for $n = 1$.[8,9] Helfrich and Müller[9] added that if the polymer is in directed rings (the monomers always adding in a certain direction), then $n = 2$. The dimension of the order parameter and the dimension of space for the system (three, for all cases discussed here) together determine the universality class for the system. Thus the nature of the polymer, whether it is in rings or in chains, is a profound issue, determining the universality class for the system. The universality class determines the functional behavior of the various thermodynamic and transport properties near the transition.

Still more interesting phenomena arise when the polymerizing compound is mixed with a solvent. Then there exist not only the polymerization transition and the usual liquid-liquid phase separations, but the possibility of an intersection of the polymerization with a phase separation to form a critical point of higher order. In 1965 Scott[10] combined the Tobolsky-Eisenberg theory of equilibrium polymerization with the Flory-Huggins theory of polymer solutions[11] and predicted the phase diagrams for mixtures of polymerizing sulfur with solvents. He predicted "ordinary" upper critical solution points at temperatures below the polymerization temperature of pure sulfur, and a second liquid-liquid phase separation, a lower critical solution point (LCSP), at temperatures above the polymerization point. This high temperature phase separation develops as follows. As solvent is added to the liquid sulfur, the polymerization point shifts to higher and higher temperatures, forming a polymerization line. That polymerization line ends at a LCSP, the presence of the polymer having engendered a phase separation. In some cases, these two regions of immiscibility were predicted to overlap. Larkin, Katz, and Scott[12] investigated experimentally ten different solvents with sulfur. Their results showed the general kinds of phase diagrams expected from Scott's theory.

In 1981, Wheeler and Pfeuty,[13] having just studied pure sulfur, as discussed above, realized that the same approach, but for the *dilute* magnet model, applies to sulfur solutions, and that Scott's theory is, naturally, the mean field limit of that model. Furthermore, he recognized that the intersection of a polymerization line with a liquid-liquid critical point can be a tricritical point, directly analogous to the tricritical point occurring in liquid mixtures of He^3 and He^4 at the intersection of the line of superfluid transitions with the liquid-liquid critical point. Kennedy and Wheeler[7] extended this approach to solutions of living polymers, in which case the polymerization transition is predicted to occur at lower and lower temperatures as the solvent is added, ending in a tricritical point at an upper critical solution point. There is, so far, no reported experimental evidence of such a phase diagram in solutions of living polymers.

We will review here our recent experimental work aimed at testing the theoretical predictions of observable behavior in sulfur and sulfur solutions. We will mention work in progress on living polymers and their solutions.

2. EXPERIMENTS ON SULFUR

The dielectric constant of liquid sulfur

Kennedy and Wheeler[14] calculated the behavior of the density of liquid sulfur near its polymerization transition. They predicted a small "dip" at the transition. Experimental measurements of Sauer and Borst[15] do show such a dip, the magnitude of which is 0.01% of the density, occurring over a temperature range of 0.5°C. The agreement between the Sauer-Borst data and the Kennedy-Wheeler theory is not bad. On the other hand, more recent data of Patel and Borst[16] show a dip in the density that is much sharper than can be explained by the theory. I was intrigued by this issue, but reluctant to try to measure again the density itself, since the effect in the density is so small.

Earlier measurements on the static dielectric constant, ϵ, of sulfur near the polymerization point showed a much bigger effect:[17, 18] a dip of 1% in ϵ, over 50°C. The static dielectric constant should reflect the same behavior as the density, amplified by the difference in atomic polarizability between monomeric and polymeric sulfur. I decided to make precise measurements of ϵ, closer to the transition temperature than had been made before.[19] Some of those measurements, together with earlier data from the literature, are shown in Figure 1. The agreement between my data and those of Baur and Horsma[18] is quite satisfactory. Also shown in Figure 1 is a theoretical line, constructed using the Clausius-Mossotti expression:

$$\frac{\epsilon - 1}{\epsilon + 2} = \frac{4\pi N_A \rho}{3M}[\phi_M \alpha_M + \phi_P \alpha_P], \tag{1}$$

where M is the molecular weight of a sulfur atom, ρ is the density (as calculated by Kennedy and Wheeler[14]), N_A is Avogadro's number, ϕ_M and ϕ_P are the mole fractions of monomer and polymer at a given temperature (as calculated by Wheeler et al.[5]), and α_M and α_P are the atomic polarizabilities of monomeric and polymeric sulfur (estimated from ϵ well below and well above the transition).

Figure 1 clearly shows that the experimental data are "rounder" than the theoretical curve. There are a number of possible causes for this difference. The Clausius-Mossotti model for ϵ could be too simplistic. The equation for the density could be too simplistic. Higher order critical terms could be important. The atomic polarizability for the polymer could depend on the extent of polymerization. Of special interest is the effect of ring

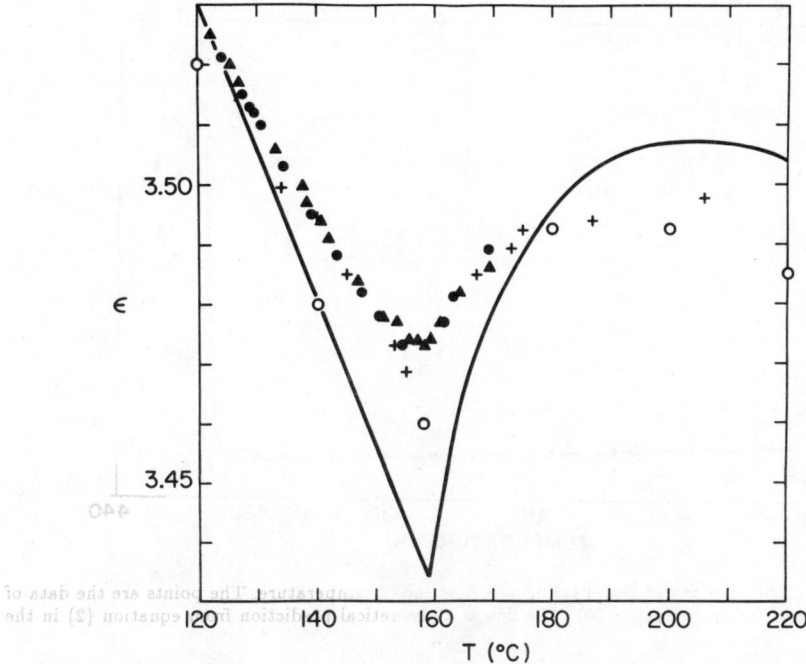

FIGURE 1. The dielectric constant of liquid sulfur as a function of temperature. The open circles are the data of Curtis (reference 17). The crosses are the data of Baur and Horsma (reference 18). The closed circles and triangles are some of the data of Greer (reference 19). The line is a theoretical prediction from equation (1) in the text.

polymers: by comparison with ring and chain hydrocarbons, we expect the presence of ring polymers to sharpen the density anomaly, but to flatten the dielectric constant anomaly. Such is just the discrepancy observed between the theory, which assumes that only chain polymers are present, and the experimental measurements of the density and the dielectric constant. Perhaps ring polymers are, indeed, present and important in liquid sulfur.

The viscosity of liquid sulfur

While using the viscosity of sulfur to determine the polymerization temperature, we made some interesting observations about the viscosity itself. The details of the technique and the data analysis will be presented in a subsequent paper.[20] We will mention here just the essential results.

First, we noted very long relaxation times for the viscosity. At temperatures just below the polymerization temperature, after a change in the temperature, a day or two was required for the viscosity to reach a steady value, during which time the value of the viscosity *increased* about 15%. At temperatures just above the transition, relaxation times were on the order of a month, during which time the viscosity *decreased* by as much as 75%! We have no good understanding of these long times. Below the transition, perhaps some polymer is already forming: the transition is not sharp. Above the transi-

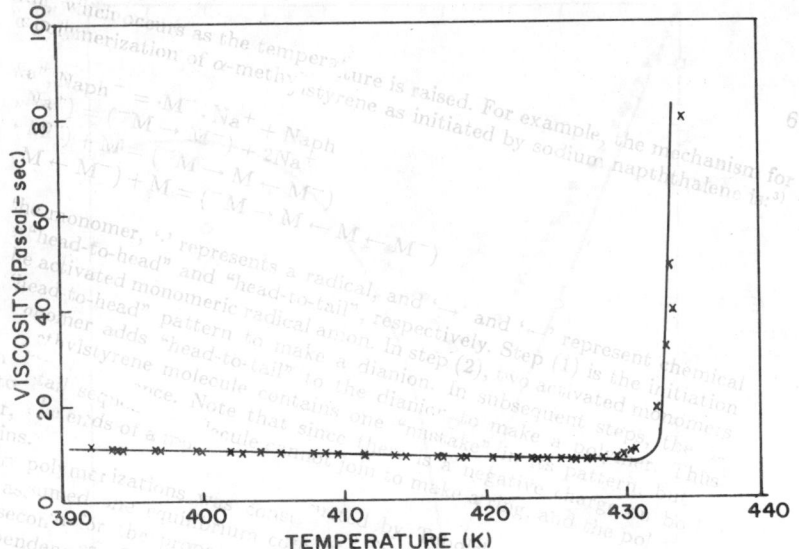

FIGURE 2. The viscosity of liquid sulfur as a function of temperature. The points are the data of Anderson and Greer (reference 20). The line is a theoretical prediction from equation (2) in the text.

tion, perhaps the polymer requires times to achieve its equilibrium distribution of chain lengths.

Figure 2 shows a plot of our viscosity measurements for pure sulfur as a function of temperature. The line is constructed from the following equation:[21]

$$\frac{\eta}{\eta_B} = 1 + [\eta]\phi(T) + k[\eta]^2\phi(T)^2 + \cdots, \tag{2}$$

where η is the viscosity at a given temperature, η_B is the background viscosity, $[\eta]$ is the intrinsic viscosity of polymeric sulfur, $\phi(T)$ is the concentration of polymer, and k is Huggins constant.[22] The η_B is determined by fitting an Arrhenius expression to the data at temperatures well below the transition. The intrinsic viscosity is taken as $BP(T)^{2/3}$, where B is a free parameter and $P(T)$ is the polymer chain length. Since $\phi(T)$ and $P(T)$ are obtained from the theory of Wheller et al.[5] and k is given the usual value of 0.69,[23] the only free parameter is B.

Figure 2 shows that the resulting equation is much "sharper" than are the data near the transition. Again, such an effect could be caused by the presence of ring polymers. On the other hand, we have to keep in mind the possible effects of non-Newtonian behavior in the viscosity, and the possibility of a critical phenomenon in the viscosity itself, an issue not yet addressed by the theorists.

The sulfur + biphenyl phase diagram

Figure 3 shows the phase diagram of sulfur in biphenyl. Included are the earlier data of Larkin, Katz, and Scott,[12] together with recent measurements made in our laboratory.[24] We had hoped to obtain more precise determinations of the phase transition lines, but were stymied because the biphenyl reacts with the diradical polymeric sulfur. Once a

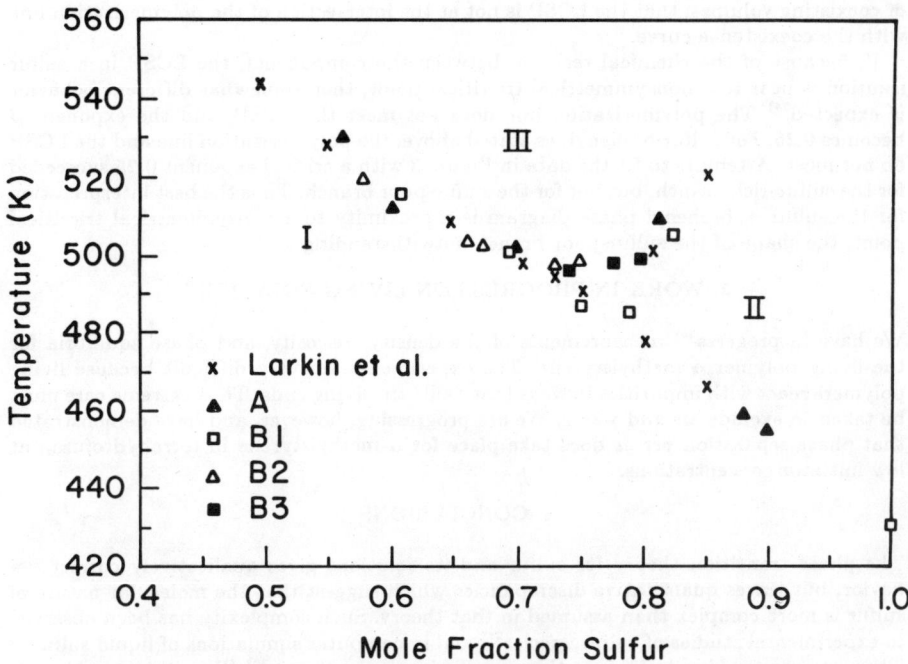

FIGURE 3. The high temperature liquid-liquid phase diagram for sulfur + biphenyl. Region I is a region of monomeric sulfur in biphenyl. Region II is a region of partially polymerized sulfur in biphenyl. Regions I and II are separated by the polymerization line. Region III is a region of coexistence of the phases of regions I and II and is defined by the coexistence curve. The key indicates which date points are from Larkin et al. (reference 12); the rest are from Anderson and Greer (reference 24).

mixture was heated through the polymerization temperature, the transition temperature began to drift constantly upward with time. Thus each datum represents observations made on a single, initial heating run for a given sample.

If the LCSP is a symmetrical tricritical point,[13] then the coexistence curve should have a critical exponent $\beta = 1$:

$$(x_1 - x_2) = Bt^\beta + \cdots = Bt + \cdots,$$

where x_1 and x_2 are the compositions of the coexisting phases, B is a constant, t is the reduced temperature $(T - T_t)/T_t$, and T_t is the tricritical (LCSP) temperature. The resulting coexisting cure will come to a sharp point at the LCSP, with the polymerization line intersecting the coexistence curve at the LCSP. From Figure 3, clearly the sulfur + biphenyl coexistence curve does not come to a point; nor did those of the other nine sulfur + solvent systems studied by Larkin et al.[12] Attempts to fit the data in Figure 3 with the predicted logarithmic correction factors[13] failed. We showed,[24] from measurements of coexisting volumes, that the LCSP is not at the intersection of the polymerization line with the coexistence curve.

If, because of the chemical reaction between the components, the LCSP in a sulfur solution is near to a nonsymmetrical tricritical point, then somewhat different behavior is expected.[25] The polymerization line does not meet the LCSP, and the exponent β becomes 0.25. For sulfur biphenyl, as stated above, the polymerization line and the LCSP do not meet. Attempts to fit the data in Figure 3 with a critical exponent 0.25 succeeded for the sulfur-rich branch, but not for the sulfur-poor branch. Thus the best interpretation for the sulfur + biphenyl phase diagram is a proximity to a nonsymmetrical tricritical point, the shape of the sulfur-poor branch notwithstanding.

3. WORK IN PROGRESS ON LIVING POLYMERS

We have in progress[26] measurements of the density, viscosity, and phase equilibria for the living polymer α-methylstyrene. The experiments are very difficult because living polymers react with impurities in ways that "kill" the living ends. Thus extreme care must be taken to exclude air and water. We are progressing, however, and have demonstrated that phase separation *per se* does take place for α-methylstyrene in tetrahydrofuran at low initiator concentrations.

4. CONCLUSIONS

The phase transition theory for sulfur and its solutions gives qualitatively correct behavior, but leaves quantitative discrepancies which suggest that the molecular nature of sulfur is more complex than assumed in that theory. Such complexity has been observed in experimental studies of sulfur species[27] and in computer simulations of liquid sulfur.[28] Theorists are working to include this complexity in the theory.[29] We will be working on further experimental tests of the theory, including measurements of the heat capacity.

It will be very interesting to see how the predictions for the properties of living polymers and their solutions compare to experiments.

REFERENCES

1. Meyer, B., *Chem. Rev.* **76**, 367 (1976).
2. Szwarc, M., *Carbanions, Living Polymers, and Electron Transfer Processes* (Interscience, New York, 1968).
3. Szwarc, M., Levy, M. and Milkowich, R., *J. Am. Chem. Soc.* **78**, 2616 (1956).
4. Tobolsky, A.V. and Eisenberg, A., *J. Am. Chem. Soc.* **81**, 780 (1959); **82** 289 (1960); *J. Colloid. Sci.* **17**, 49 (1962).
5. Wheeler, J.C., Kennedy, S.J. and Pfeuty, P., *Phys. Rev. Lett.* **45**, 1748 (1980).
6. Wheeler, J.C. and Pfeuty, P., *Phys. Rev.* **A24**, 1050 (1981).
7. Kennedy, S.J. and Wheeler, J.C., *J. Chem. Phys.* **78**, 953 (1983).
8. Cordery, R., *Phys. Rev. Lett* **47**, 457 (1981); Pfeuty, P. and Wheeler, J.C., *Phys. Lett.* **84A**, 493 (1981); Duplantier, B. and Pfeuty, P., *J. Phys.* **A15**, L127 (1982); Gujrati, P.D., *Phys. Rev.* **B27**, 4507 (1983); Helfrich, W., *J. Phys. (Paris)* **44**, 13 (1983).
9. Helfrich, W. and Müller, W., *Continuous Models of Discrete Systems* (University of Waterloo Press, Waterloo, Ontario, 1980), p. 753.
10. Scott, R.L., *J. Phys. Chem.* **69**, 261 (1965).
11. Flory, P.J., *J. Chem. Phys.* **10**, 51 (1942); Huggins, M.L., *Ann. N.Y. Acad. Sci.* **43**, 1 (1942).
12. Larkin, J.A., Katz, J. and Scott, R.L., *J. Phys. Chem.* **71**, 352 (1967).

13. Wheeler, J.C. and Pfeuty, P., *Phys. Rev. Lett* **16**, 1409 (1981); *J. Chem. Phys.* **74**, 6415 (1981).
14. Kennedy, S.J. and Wheeler, J.C., *J. Chem. Phys.* **78**, 1523 (1983).
15. Sauer, G.E. and Borst, L.B., *Science* **158**, 1567 (1967).
16. Patel, H. and Borst, L.B., *J. Chem. Phys.* **54**, 822 (1971).
17. Curtis, H.J., *J. Chem. Phys.* **1**, 160 (1933).
18. Baur, M.E. and Horsma, D.A., *J. Phys. Chem.* **78**, 1670 (1974).
19. Greer, S.C., *J. Chem. Phys.* **84**, 6984 (1986).
20. Anderson, E.M., Ph.D. Dissertation, The University of Maryland at College Park, 1987; Anderson, E.M. and Greer, S.C., work in progress.
21. Gee, G., *Trans. Faraday Soc.* **48**, 515 (1952).
22. Huggins, M.L., *J. Am. Chem. Soc.* **64**, 2716 (1942).
23. Peterson, J. and Fixman, M.J., *J. Chem. Phys.* **39**, 2516 (1963).
24. Anderson, E.M. and Greer, S.C., *J. Chem. Phys.* (in press).
25. Wheeler, J.C., *Phys. Rev. Lett.* **53**, 174 (1984); Wheeler, J.C., *J. Chem. Phys.* **81**, 3635 (1984).
26. Ruiz-Garcia, J., Zheng, K.M. and Greer, S.C., work in progress.
27. Steudel, R., *Studies in Inorganic Chemistry 5: Sulfur, its Significance for Chemistry, for the Geo-, Bio-, and Cosmophere and Technology*, edited by A. Müller and B. Krebs (Elsevier, New York, 1984), p. 2–37.
28. Stillinger, F., Weber, T.A. and LaViolette, R.A., *J. Chem. Phys.* **85**, 6460 (1986).
29. Wheeler, J.C., Petchek, R.G. and Pfeuty, P., *Phys. Rev. Lett.* **50**, 1633 (1983); Petschek, R.G., *Phys. Rev.* **A34**, 2391 (1986).

INTERACTION BETWEEN POLYMERS AND COLLOIDAL PARTICLES

P. Pincus

Materials Department, University of California, Santa Barbara,
CA 93106

ABSTRACT. We review the current status of colloid stabilization control by adsorbed homopolymers in non-polar solvents. In particular, a set of rules which are necessary to achieve colloid stabilization are presented. The basic physics underlying the rules is discussed in the context of both flat surfaces and small colloidal particles.

1. INTRODUCTION

The control of the stability of colloidal dispersions of solid particles in liquids plays an important role in many technologies ranging from powder processing of ceramics to water treatment. For several applications, it is insufficient to simply stabilize the suspensions but, in addition, one would like to indicate the structure of the destabilized species, *e.g.* weak flocs, dense aggregates, gels, etc. The understanding of the relevant control conditions depend on a variety of parameters such as particle diffusion constants, densities, chemical reactivity etc., but also, very sensitively on the nature of the interparticle forces. It is this latter area which shall be the focus of this presentation.

The central problem in achieving colloid stabilization is posed by the ubiquitous Van Der Waals attractive interactions[1] between like objects. If two spherical colloidal particles of radius b, have a distance of closest approach δ, ($\delta \ll b$) this interaction is given by

$$U_{vw} \simeq -A(b/\delta) \qquad (1)$$

where the Hamaker constant A depends upon the contrast in dielectric function between the solid and fluid. Typically $A \sim (10^{-1} - 10)T_0$, where T_0 is the ambient temperature in units of Boltzmann's constant. Thus, if two Brownian particles come into contact, they become bound by an arbitrarily large energy leading to *irreversible* sticking. Continuation of this process may lead to eventual sedimentation or creaming. It is precisely this dispersion force driven "phase separation" which we would like to control.

The methods used to achieve colloid stabilization depend upon the polarity of the solvent. In polar solvents, such as water and some alcohols, which support charge separation, ionic groups may be chemically bonded to the surfaces of particles. Then, upon dissociation charged particles result which interact with one another via a screened Coulomb repulsion. This is the case for emulsion polymerized lattices[2] and micelles self-assembled from ionic surfactants[3] While, in detail, the interaction between double layers, especially at short distances, is imperfectly understood,[4,5] at longer distance modelling by a Yukawa potential[6] seems reasonable. At low ionic strengths, this interaction, may be sufficiently strong for large particles[6] to lead to ordered colloidal crystal structures[2] At higher ionic strengths, *i.e.* with added salts, Debye screening reduces the range of the interaction and may lead to eventual flocculation —this is the "salting out" effect.

For ceramic processing[7] in aqueous solvents, it is more common to use charged polymers, polyelectrolytes, such as sulfonated polystyrene or polyacrylic acid. If these polymers physisorb onto the colloidal particles stabilization may be achieved by a combination of charge and entropic (to be discussed later) effects. The advantage here is that by relatively mild treatment the polymer may be eventually removed. While some theoretical effort[8,9] has gone into studying this situation, the problem is exacerbated by our relatively poor understanding of the properties of bulk polyelectrolyte solutions[10-13]

In non-polar solvents such as cyclohexane, toluene, etc., where charge separation is difficult, colloidal stability is achieved either through steric stabilization with adsorbed polymers or chemical grafting of polymers which are repelled from the particulate surface. This review will be mainly concerned with steric stabilization by homopolymers.

For the case of grafted chains, however, the stabilization mechanism is reasonably clear. If a relative high density of polymers, which do *not* wet the solid as well as does the solvent, are covalently bonded to the surface (typically at one end of the chain), they will make a corona extending into the solvent a distance L. Scaling relations have been developed for the corona associated with flat surfaces, L_0, by Alexander[14] and de Gennes[15] They find

$$L_0 \simeq N(a/D)^{1/\nu}D \qquad (2)$$

where N is the number of statistical segments of size a which comprises the chain and D is the mean distance between grafting sites. For good solvents the exponent[20] $\nu \simeq 0.6$. The linearity of L_0 as a function of N is indicative of stretched chains. More recently Hirz[16] and Witten *et al.*[17-18] have given detailed concentration profiles for this case. This behavior obtains for spherical particles until the particulate radius, b, becomes comparable to L_0. For $b < L_0$, Daoud and Cotton[19] have shown that the corona thickness $L(b)$ scales as N^ν, as for free chains. Writing $L = L_0 f(L_0/b)$, where $f(x)$ is a scaling function such that $f(0) = 1$ and $f(x) \propto x^p$ for $x \gg 1$, we find

$$L \simeq N^\nu a(b/D)^{1-\nu} \qquad (3)$$

which is the Daoud-Cotton result. Clearly, under such circumstances, the interaction between two coronae is purely repulsive[17-18] Then, to a reasonable approximation, if

$$A(b/L) < T \qquad (4)$$

stabilization against Van der Waals induced segregation is achieved. This generally requires $L \sim b$ and is most efficiently obtained with a high grafting density of short chains[21]

Our major concern here is with the more delicate situation that arises when stabilization control is managed with physisorbed homopolymers. Over the last few years, a significant effort, both theoretical and experimental, has been expended in this area. Integrating the results of several groups, we are led to postulate the following set of rules which must be satisfied for stabilization (against Van Der Waals interaction induced aggregation):

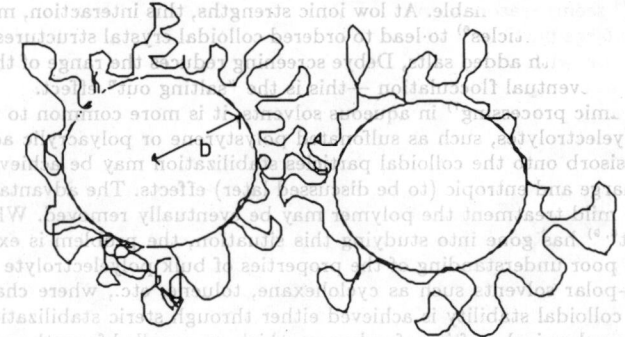

FIGURE 1. A sketch of two interacting colloidal particles with adsorbed polymer. Note the deformation of the polymer distribution in the interstitial region.

i) The colloidal particle must provide an *adsorbing* substrate for the polymer/solvent couple.
ii) The solvent should be *good* for the polymer.
iii) The Polymer should be *flexible*.
iv) The molecular weight of the polymer should be sufficiently high that the adsorption is irreversible over the relevant experimental time scales.
v) There should be sufficient polymer to *saturate* the surfaces.
vi) The polymer corona must be sufficiently *thick* so that a condition such as (4) is obeyed to prevent Van der Waals "sticking" of the particles.

The remainder of this paper will discuss the physical principles which underlie these rules. The following section will specifically deal with flat parallel surfaces which provide a fairly simple model for the contacting area of two colloidal particles. In section 3, these ideas will be extended to small particles where specific finite surface area aspects may be important.

The qualitative picture (Figure 1) which provides the basis of steric stabilization relies on the adsorption of polymers onto the colloidal particles in a series of loops and trains[23] that extend a finite corona thickness, L, into the solvent. Then, during a collision between a pair of particles, the region between the two surfaces has a concentration of polymer which is increasing as the particles approach one another. For good solvents, this results in an increased positive osmotic pressure which acts as a "bumper" and tends to keep the particles separated. If the polymer was repelled by the surface, the intergranular space would then be depleted of polymer resulting in an additional attractive force between the particles arising from the uncompensated osmotic pressure associated with the dissolved polymer chains outside the interstitial region. This is the attractive depletion force[22] Thus the rationale behind the rules i-iii is quite clear even within this simple cartoon model. In the next section, we shall make the geometry of the adsorbed layer more precise and introduce the idea of bridging which leads to some subtleties.

2. FLAT SURFACES

Let us first consider the geometry of the adsorption of a single chain on a flat surface (Figure 2). De Gennes[20] has given a simple argument for the thickness, D, of the resulting "pancake" as determined from a balance between adsorption and chain elasticity. The

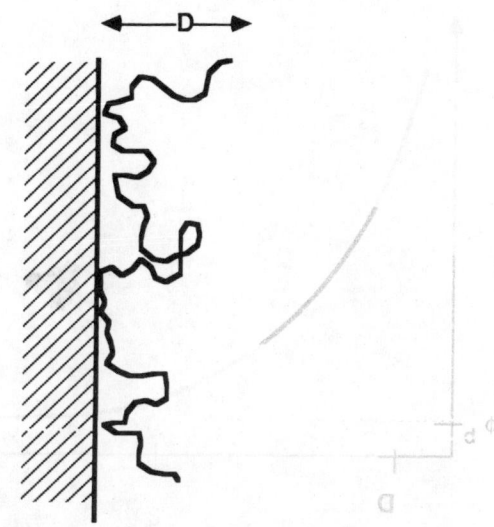

FIGURE 2. Sketch of a polymer chain adsorbing on a flat surface. Note that the coil is constrained in the direction normal to the surface and spreads out along the surface.

free energy, F, of an adsorbed polymer may be approximately expressed as

$$F/T \simeq (R_0/D)^2 - \epsilon N x \qquad (5)$$

where $R_0^2 = Na^2$ is the characteristic size of a random walk chain composed of N statistical segments of length a; ϵT is the adsorption energy gained per segment in contact with the surface; x is the fraction of segments which contact the surface. The first term represents the reduction of the chain entropy associated with the confinement of the polymer to the slab of width D; the second term is the enthalpic gain when monomers touch the surface. Making the simplest geometrical ansatz that $x \simeq (a/D)$ and minimizing F with respect to D, we obtain

$$a/D \simeq \epsilon \qquad (6)$$

Thus, provided that $\epsilon > N^{-1/2}$, the polymer is strongly deformed. Typically $\epsilon \approx 1$, since it generally also arises from dispersion forces. Thus this inequality is easily obeyed for polymers with $N \geq 100$. In the presence of excluded volume interactions, each term in F is modified[24] but the result (6) for D is unaltered.[25] For ideal chains, the adsorption-free energy per chain is of order $N\epsilon^2 T$, which is generally much greater than T. This implies a high barrier against chain desorption by thermal fluctuations. As we shall see later, this is a crucial point required for colloid stabilization.

For a flat adsorbing surface in equilibrium with a bulk polymer solution, even if it is very dilute, the surface region contains many overlapping chains, i.e., it is semi-dilute. This occurs because of the relatively small entropy of mixing for high molecular weight polymers. Bouchaud and Daoud[26] have studied in detail ultra-dilute solutions where incomplete surface coverage occurs. An appropriate function to describe the adsorption is then the concentration profile, $\varphi(z)$, where z is the distance from the surface and $\varphi(z)$

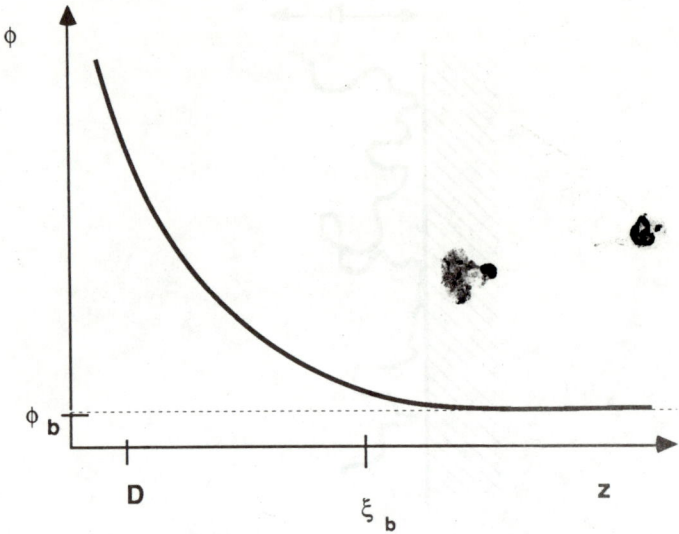

FIGURE 3. A sketch of the concentration profile for an interface in contact with a semi-dilute polymer solution. The profile extends into the solvent a characteristic distance ξ_b and approaches the bulk concentration exponentially for larger distances.

is the local monomer volume fraction. De Gennes[27] has suggested that in good solvents, this monotonic profile may be broken down into three regions (Figure 3): proximal $z < D$; central $\xi_b > z > D$, distal $z > \xi_b$ where D is the single adsorbed chain thickness (6) and ξ_b is the bulk correlation length. For a bulk semi-dilute polymer solution, the correlation length ξ_b, which approximately corresponds to the mesh size of the transient network,[20] is a monotonically decreasing function of the bulk volume fraction φ_b,

$$\xi \simeq a\varphi^{-p}; \qquad p = \nu(d\nu - 1) \qquad (7)$$

where $p = 3/4$ good solvents. At the overlap concentration, ξ crosses over to the Flory radius, $R_F = N^\nu a$ ($\nu = 3/5$). In the proximal regime $z < D$, a new surface exponent, the proximal exponent[24, 25] comes into play. However, for strong adsorption, $\epsilon > 1$, D becomes comparable to a monomer dimension and this region is quite small. Nevertheless, it is the proximal region which sets the amplitude scale for the deviation of the local polymer concentration from its bulk value. The central regime, which usually dominates the physics, displays a power law decay for the concentration which is an example of self-similarity. The relevant exponent for the decay of the profile is p^{-1} which may easily be seen as follows: The profile must be a function of a dimensionless length, i.e., $\varphi(z/\lambda)$ where λ is some characteristic length. What is λ? If we employ the basic tenet of polymer scaling theory that the only relevant length is ξ, we are tempted to identify λ with ξ. But, since ξ is a function of concentration, it diminishes as we approach the surfaces, i.e., $\xi = \xi(\varphi)$. These remarks can be consistent only if $\xi(\varphi) \propto z$. Then, using (7), we find

$$\varphi(z) \simeq \varphi_s(D/z)^{1/p} \qquad (8)$$

in the central region, where φ_s is the amplitude in the proximal region. Evidence for the unusual 4/3 power law decay has been provided by neutron scattering from colloidal particles with adsorbed polymer[28] and neutron reflectometry[29] from free liquid surfaces. In mean field theory, the decay is parabolic, $p = 1/2$. In the distal region, $z > \xi_b$, the profile decays exponentially as $e^{-(z/\xi_b)}$ toward φ_b. The corona thickness L is then effectively ξ_b for semidilute solutions and saturates at R_F for dilute solutions (so long as we avoid the ultra-dilute Daoud[26] regime). Note that in semi-dilute solutions, L *decreases* as the concentration increases. This is the screening effect of interchain interactions.

What happens when two such flat surfaces are pushed together? As the profiles overlap, there are two opposing effects: the proximity of the two surfaces provides more surface area on which a given chain may adsorb, *i.e.*, *bridging* induced attraction; the previously mentioned osmotic repulsion arising from increasing polymer concentration in the interstitial channel, *i.e. entropic* repulsion. Which contribution dominates? In thermodynamic equilibrium, it is the *bridging attractive* force.[30] The underlying physics is a generalization of the Van Der Waals Theorem that the dispersion force between two identical particles is attractive. Imagine inserting a particle into a deformable medium. The particle locally perturbs the medium which responds over a length scale, ξ. The injection of the particle requires a work associated with the deformation created. If there is no direct interaction between the particles, a second injected particle will preferentially move to the vicinity of the first so as to take advantage of the already existing deformation. Thus, the work required for its injection will be diminished if it is found within a correlation distance of the first particle.

A more analytic, albeit approximate, method to yield the same result of attractive interactions in thermodynamic equilibrium is given by Cahn-de Gennes polymer adsorption theory. Then, if the profile $\varphi(z)$, is treated as a variable, the interfacial free energy is given by[27, 30]

$$\frac{\gamma}{2} = \gamma_1(\varphi_s) + \int_0^h \left[F\{\varphi\} - \mu\varphi + \pi + mT\left(\frac{d\varphi}{dz}\right)^2 \right] dz \qquad (9)$$

where $\gamma_1(\varphi_s)$ describes the local interaction between the polymer and the surface [equivalent to the ϵ term in (5)]; $2h$ is the separation of the surfaces; $F\{\varphi\}$ is the bulk Helmholtz free energy functional; μ the chemical potential; π is the osmotic pressure and the square gradient term is related to the chain elasticity. The profile is determined (if F, μ and π are known) by deriving en Euler-Lagrange equation by minimizing γ with respect to $\varphi(z)$. The disjoining pressure, P between the surfaces is then simply

$$P = -\left(\frac{\partial \gamma}{\partial h}\right) = \mu\varphi_h - F\{\varphi_h\} - \pi = -\Omega(\varphi_h) \qquad (10)$$

where φ_h is the volume fraction on the symmetry plane and the direct contribution from the square gradient term vanishes by symmetry; $\Omega = F - \mu\varphi + \pi$ is the grand potential density. In the bulk of the solution, $\Omega(\varphi_b) = 0$, and mechanical stability requires $\Omega(\varphi) \geq 0$. Thus $\Omega(\varphi_h)$ is positive definite, implying an attractive force between the surfaces.

The essential reason that thermodynamic equilibrium dictates attractive forces between the surfaces depends upon the mechanism for repulsions being the increased osmotic pressure associated with increased interstitial polymer concentration. This implies an *increased* free energy for those chains between the surfaces relative to those in the bulk solution. Then, in the absence of any other constraints, these "unhappy" polymers could desorb and diffuse into the reservoir. This process will continue until the bridging attraction dominates the situation.

How, then, are repulsive forces achieved? If, during the relevant experimental time scale of surface approach, the interstitial polymers do *not* have sufficient force to equilibrate with the bulk solutions, repulsions should be favored over the pure equilibrium case. With this in mind, we must consider the history of any experimental process. In this regard, several groups[31] have effectively measured the force between closely approaching surfaces mediated by a polymer solution in the range $10^4 \text{ Å} > h > 1 \text{ Å}$. The typical procedure would be to incubate the two surfaces in the solution for several hours at very large separations —this allows the adsorption to approach thermal equilibrium. The surfaces are then brought together and the forces measured mechanically. Because of the typical strong barrier against chain description (e^{-N^2}), de Gennes[30] suggested that a reasonable scenario is that polymers could not effectively desorb during the measurement time. Then, rather than thermal equilibrium, the ansatz of a constrained equilibrium with the surface excess,

$$\Gamma \equiv \int_0^\infty (\varphi - \varphi_b) dz$$

held fixed, might be more appropriate to describe the experiments. Yet, even with this constrained equilibrium preventing the chains from scaping from between the surfaces, the *sign* of the force is not necessarily positive.

Referring to (9), the simplest assumption about the local coupling between the polymer and the solid surface which is equivalent to (5) is

$$\gamma_1(\varphi_s) = -\epsilon \varphi_s T \tag{11}$$

where ϵ has the same meaning as in (5). Equation (9) may easily be modified to include the constraint by replacing the chemical potential μ by Lagrange multiplier, the pseudo-chemical potential $\tilde{\mu}$ such that the constraint is obeyed at any separation, $2h$. In mean field Flory-Huggins theory where $p = 1/2$ and

$$F(\varphi) = \frac{T}{2} v \varphi^2;$$

$$\pi = \left[\varphi \left(\frac{\partial F}{\partial \varphi} \right) - F \right]\bigg|_b = \frac{T}{2} v \varphi_b^2; \tag{12}$$

$$m(\varphi) = m_0 \varphi$$

together with (11), De Gennes[30] has shown that there is an *exact* cancellation between the bridging and osmotic effects leading to *zero* net force. Thus, we see that even the *sign* of the force is quite delicate. However, in an improved scaling model with $p = 3/4$ and

$$F(\varphi) \simeq T \xi^{-3}(\varphi); \quad \pi \simeq T \xi^{-3}(\varphi_b) \tag{13}$$

a purely *repulsive* force is found.[30] Indeed, when the interplane distance corresponds to the central regime, we expect a power law force which scales as

$$P \simeq T h^{-3} \tag{14}$$

since the disjoining pressure must have dimensions of energy density with T giving the energy scale and the only relevant length being h. In the distal region, we expect $P \sim$

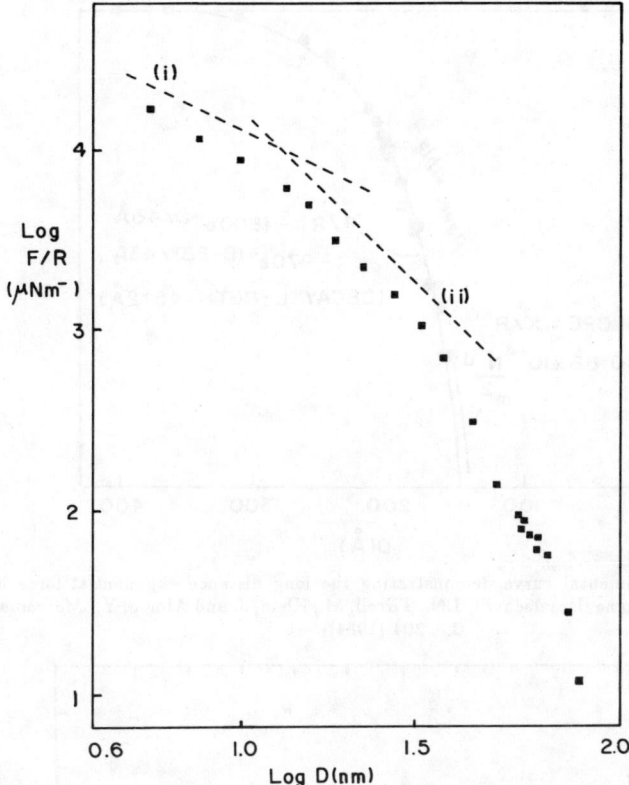

FIGURE 4. Experimental curve of the force versus separation between two mica surfaces immersed in an aqueous polyethylene oxide solution [Klein, J. and Luckham, P.F., *Macromolecules* **17**, 1041 (1984)]. Since water is a good solvent, note the purely repulsive force.

$e^{-(z/\xi_b)}$. Indeed, these behaviors are consistent with existing data in good solvents,[31] (Figures 4 and 5).

In summary, for constrained equilibrium with irreversibly adsorbed chains and in good solvents, a repulsive disjoining pressure is predicted. The thickness of the corona is ξ_b with a dilute solution maximum of R_F. Provided that $A \leq T$, this should be sufficient to give a *net* repulsive force even in the presence of Van der Waals attractions between the solid surfaces. As might be expected, when the solvent quality is lowered to the theta[32] or poor solvent regions,[33, 34] attractive forces appear once again for certain ranges of separation (Figures 6 and 7). In fact, even in good solvents under special conditions, attractive forces may develop. This occurs[35] with "starved" surfaces where the constrained surfaces excess Γ is less than the incubated thermal equilibrium value Γ_{eq}. Not only does the theory predict that an equilibrium position develops as Γ decreases, but these results are in qualitative agreement with experiments with incomplete incubation[36] (Figure 8) and adsorption sterically hindered during incubation.[37] The reason for the attractions to develop at large separations is again that as the total amount of polymer diminishes between the surfaces, the relative weight of bridging to osmotic pressure increases. Indeed,

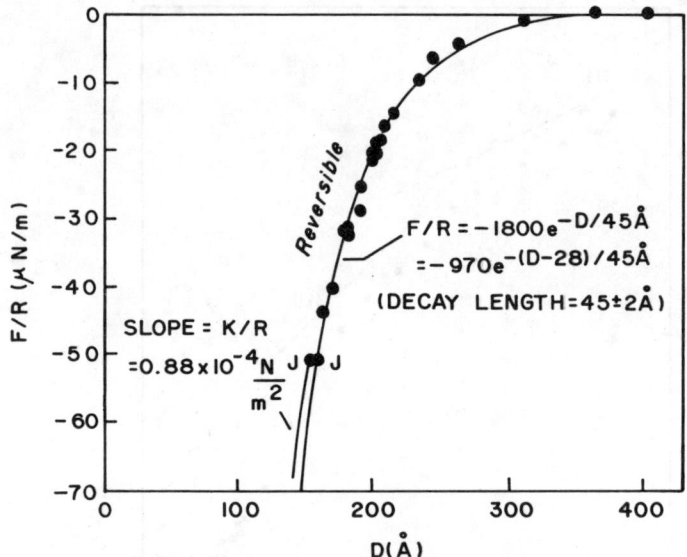

FIGURE 5. An experimental curve demonstrating the long distance exponential force law for polystyrene in cyclohexane [Israelachvili, J.N., Tirrell, M., Klein, J. and Almog, Y., *Macromolecules* **17**, 204 (1984)].

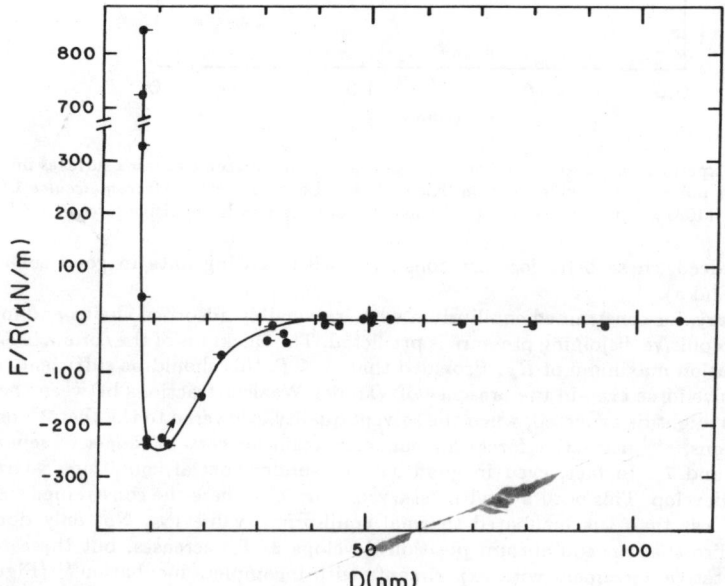

FIGURE 6. Force versus distance curve for PS in cyclopentane, a theta solvent [Almog, Y. and Klein, J., *J. Colloid and Interf. Sci.* **106**, 33 (1985)].

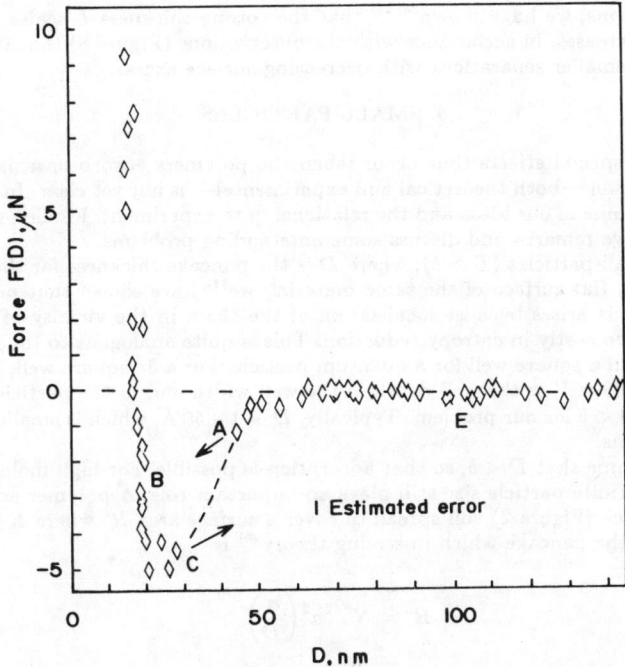

FIGURE 7. PS in cyclohexane below the Theta temperature in poor solvent conditions [Israelachvili, J.N., Tirrell, M., Klein, J. and Almog, Y., *Macromolecules* **17**, 204 (1984)].

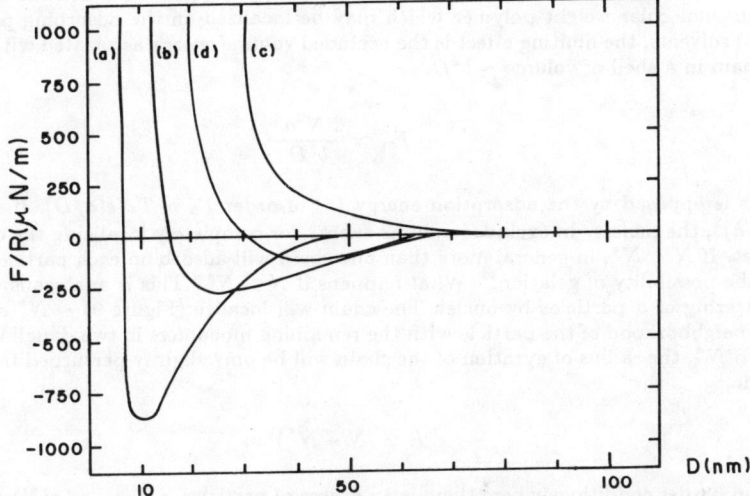

FIGURE 8. Development of the force versus distance curve as a function in increasing incubation times. Note that the starved surfaces have an adhesive minimum while the saturated surfaces are monotonically repulsive [Almog, Y. and Klein, J., *J. Colloid and Interf. Sci.* **106**, 33 (1985)].

in dilute solutions, we have shown[32,35] that the corona thickness L scales between R_F and D as Γ decreases, in accordance with the observations (Figure 8) that the minimum moves toward smaller separations with decreasing surface excess.

3. SMALL PARTICLES

What are the special effects that occur when the polymers adsorb on small particles? Here the situation —both theoretical and experimental— is not yet clear. In this section, I will present some of our ideas and the relationship to experiment. Finally, we will make some speculative remarks and discuss some outstanding problems.

For very small particles ($D > b$), where D is the pancake thickness for the adsorption of a chain on a flat surface of the same material, we[38] have shown that *no* adsorption is expected. This arises because localization of the chain in the vicinity of a point-like object is just too costly in entropy reduction. This is quite analogous to the formation of a bound state in a square well for a quantum particle. For a $3d$ square well, this requires $Va^2 \geq \hbar^2/m$ where V is the well depth, a the well width and m the particle mass. This translates to $D < b$ for our problem. Typically, $D \sim 1 - 50$ Å, which is smaller than most colloidal systems.

We now assume that $D < b$, so that adsorption is possible. For high molecular weight polymers, the finite particle size still plays an important role. A polymer adsorbed on a large flat surface (Figure 2) will spread to cover a surface area R^2 where R is the radius of gyration of the pancake which in scaling theory[20] is

$$R^2 = N^{3/2} a^2 \left(\frac{a}{D}\right)^{1/2} \tag{15}$$

where the molecular weight exponent is characteristic of polymers in good solvent confined to two dimensions. Clearly, for a finite particle of surface area proportional to b^2, there is not sufficient adsorption sites for a high molecular weight chain. An elementary estimate[38] may be made of the adsorbance capacity, N^*, which is defined as the maximum molecular weight polymer which may be localized on the adsorbing particle. For good solvents, the limiting effect is the excluded volume energy associated with localizing a chain in a shell of volume $\sim b^2 D$,

$$F_{ev} \simeq \frac{TN^2 a^3}{b^2 D}. \tag{16}$$

This is opposed by the adsorption energy (5) of order $F_a \simeq TN\epsilon(a/D)$. If $N > N^* \simeq \epsilon(b/a)^2$, the penalty in excluded volume energy for completely localizing the chain is too great. If $N < N^*$, in general more than one chain will adsorb on each particle[39] leading to the possibility of gelation.[40] What happens if $N > N^*$? This is analogous to resonant scattering of α particles by nuclei. The chain will localize (Figure 9) $\sim N^*$ segments in the neighborhood of the particle with the remaining monomers in two dangling ends. For $N \gg N^*$, the radius of gyration of the chain will be only slightly perturbed from its bulk value,

$$R \simeq (N - N^*)^\nu a. \tag{17}$$

Then, under conditions where there is an excess of particles, *i.e.*, $c_p > (c/N)$, where c_p is the concentration of colloidal grains, several particles may adsorb on *each* strand forming

Cabane-Duplessix necklaces (Figure 10). Indeed, these authors[41] have observed **precisely** this type of structure in aqueous solutions of polyethylene oxide with SDS micelles.

Returning now to the case of the "one pearl necklace" (Figure 9), we may consider the polymer contribution to the interaction between two such "hairy" beads. This problem is isomorphic to the interaction between particles each of which has two grafted chains.[21] In this case, a scaling discussion gives an unusual logarithmic potential of mean force between the particles ($N \gg N^*$) and ($R \gg b$),

$$\frac{U(r)}{T} \simeq \Theta \ln\left(\frac{R}{r}\right) \qquad (18)$$

where r is the separation between the centers of the grains and R is the chain radius (17); $\Theta \simeq 0.8$ is a DesCliseaux exponent describing correlations between *internal* segments on an excluded volume chain. Under these conditions, effective stabilization is achieved if

$$\frac{A}{T} \leq \Theta(R/b)^6 \qquad (19)$$

which is easily obeyed for $R > b$.

What is the effect of excess polymer? As we expect from the discussion in Section 2 and has been recently demonstrated,[39,43] the corona thickness of the single beaded strands diminishes from the radius given in (17) to the bulk correlation length ξ_b. The cut-off in (18) then becomes ξ_b instead of R tending to *destabilize* the suspension, when the inequality (19) (with ξ_b replacing R) is no longer satisfied. Note, however, that this generally occurs for $\xi \leq b$ where particulate diffusion is quite slow.[44] Then, even a thermodynamically unstable phase may take a long time to segregate.

Returning now to the case of excess colloidal grains, we have already seen that when $N \gg N^*$, Cabane-Duplessix necklaces (Figure 10) are predicted and observed. How do the necklaces interact with one another? As a necklace is increasingly loaded with beads, the effective solvent quality for the necklace diminishes, *i.e.*, the average interchain interactions are *less* repulsive. This effect may be simply described in terms of a virial theory. To a first approximation, the repulsive polymer induced interaction between beads will yield a fairly uniform spacing of beads along the necklace. Then we may renormalize the chain containing n beads as containing n segments which each look like Figure 9. Then the potential of mean force between two such segments with centers of gravity separated by r is

$$V(r) = V_{vw}(r) + U(r) \qquad (20)$$

where $U(r)$ (18) is the polymer contribution and $V_{vw}(r)$ is the dispersion interaction between the bare colloidal grains. The usual expression for the second virial coefficient (or excluded volume)

$$v = \int \{1 - e^{-[V(r)/T]}\} d^3r \qquad (21)$$

becomes

$$v = \int \left\{1 - \left(\frac{r}{\lambda}\right)^\theta e^{-[V_{vw}(r)/T]}\right\} d^3r \qquad (22)$$

where $\lambda \simeq m^\nu a$; $m = (N/n) - N^*$ is the mean number of monomers between beads. So

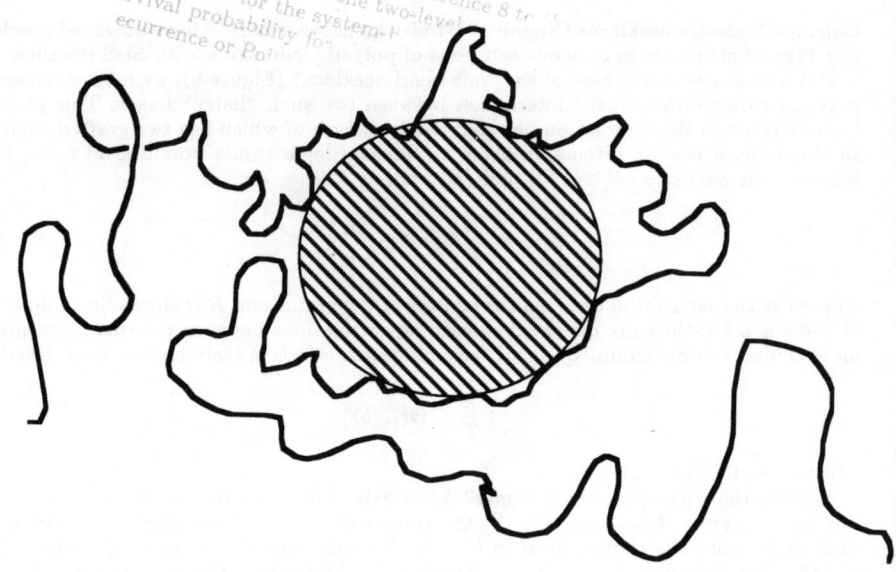

FIGURE 9.

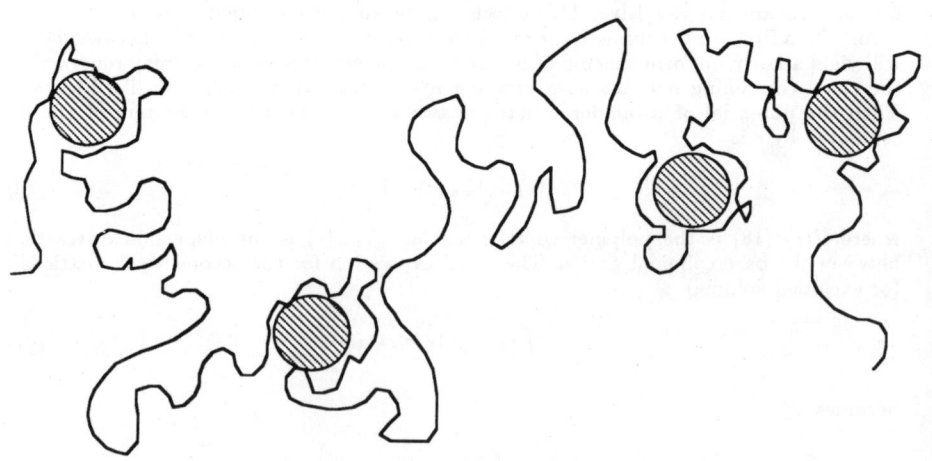

FIGURE 10.

long as $v > 0$, the necklaces are effectively in good to marginal solvents. Clearly, as n increases, any attractive dispersion force tends to destabilize the solution of necklaces. The size of one "hairy bead" crosses over from $\lambda \simeq m^\nu a$ for $m > (b/a)^{1/\nu}$ to $\lambda \simeq b$ when the inequality is not satisfied. Then a simple Flory calculation yields,

$$R = n^{3/5} \lambda \left(\frac{v}{\lambda^3}\right)^{1/5} \tag{23}$$

where the excluded volume per hairy bead v is given by (22). When $v < 0$, the necklace collapses and we have segregation.

For $N \leq N^*$, there is an important question[39] which has yet to be completely solved. This is the important problem of how many polymer chains attach to each colloidal particle and what is the associated distribution function? If this mean number exceeds two, the beads may act as cross-links, ultimately yielding a physical gel. How is the competition between such gelation and a simple segregation of the particles described? Can gels form with $N > N^*$? These are some issues which are both scientifically challenging and technologically important.

ACKNOWLEDGEMENTS

I would like to thank many colleagues for their collaboration and discussions on polymer at surfaces, including T. Witten, P.G. de Gennes, J.F. Joanny, B. Cabane, J. Klein, Hong Ji, G. Rossi, D. Hone, K. Kremer, and G. Hadziioannou. Part of this paper was written at the Physics Institute, Mainz, with the kind hospitality of K. Binder and K. Kremer and financial support of a NATO Travel Grant 86/680. This work has been partially sponsored by a U.S. Department of Energy grant to UCSB on "Polymers at Surfaces".

REFERENCES

1. Mohanty, J. and Ninham, B.W., *Dispersion Forces* (Academic Press, New York, 1976).
2. "Les Houches Meeting", *J. de Physique* **46**, C3 (1985).
3. Israelachvili, J.N., *Intermolecular and Surface Forces: With Applications to Colloidal and Biological Systems* (Academic Press, New York, 1985).
4. Wennerstrom, J., *J. Chem. Phys.* **79**, 2221 (1984).
5. Kjellander, R. and Marcelja, S., *Chem. Phys. Let.* **127**, 402 (1986); Kjellander, R. and Marcelja, S., *J. Phys. Chem.* **90**, 1230 (1986).
6. Alexander, S., Chaikin, P.M., Grant, P., Morales, G.J., Pincus, P. and Hone, D., *J. Chem. Phys.* **80**, 5776 (1984).
7. Fennelly, T.J. and Reed, J.S., *J. Am. Ceram. Soc.* **55**, 264 (1972).
8. Barford, W., Ball, R.C., Nex, C.M.M., *J. Chem. Soc. Faraday Trans.* **82**, 3233 (1986).
9. van der Schee, H.A. and Lyklema, J. (to be published).
10. de Gennes, P.G., Pincus, P., Velasco, R.M. and Brochard, F., *J. Phys. (Paris)* **37**, 1461 (1976).
11. Odijk, T., *Macromolecules* **12**, 688 (1979).
12. Mandel, M. To appear in *Encyclopedia of Polymer Science and Engineering*, Second Edition (John Wiley & Sons, New York).
13. Witten, T.A. and Pincus, P., *Europhysics Letters* **3**, 315 (1987).
14. Alexander, S, *J. Physique* **38**, 983 (1977).
15. de Gennes, P.G., *Macromolecules* **13**, 1069 (1980).
16. Hirz, S., Thesis, Chemical Engineering, University of Minnesota (unpublished).
17. Milner, S.T., Witten, T.A. and Cates, M.E., *Macromolecules* (to be published).

18. Milner, S.T., Witten, T.A. and Cates, M.E., *Europhysics Lett.* **5**, 413 (1988).
19. Daoud, M. and Cotton, J.P., *J. de Physique (Paris)* **43**, 531 (1982).
20. de Gennes, P.G., *Scalling Concepts in Polymer Physics* (Cornell Univesity Press, Ithaca, 1979).
21. Witten, T.A. and Pincus, P.A., *Macromolecules* **19**, 2509 (1986).
22. Napper, D.H., *Polymeric Stabilization of Colloidal Dispersions* (Academic Press, London, 1983).
23. Fleer, G.J. and Scheutjens, J.M.H.M., *Adv. Colloid Interface Sci.* **16**, 361 (1982).
24. Eisenriegler, E., Kremer, K. and Binder, K., *J. Chem. Phys.* **77**, 6296 (1982).
25. de Gennes, P.G. and Pincus, P., *J. Physique-Lettres* **45**, L953 (1984).
26. Bouchaud, E. and Daoud, M., *J. de Physique* **48**, 1991 (1987).
27. de Gennes, P.G., *Macromolecules* **14**, 1637 (1981).
28. Auvray, L. and Cotton, J.P., *Macromolecules* **20**, 202 (1987).
29. Bouchaud, E., Farnoux, B., Sun, X., Daoud, M. and Jannink, G., *Europhys. Let.* **2**, 315 (1986); Sun, X., Bouchaud, E., Lapp, A., Farnoux, B., Daoud, M. and Jannink, G. (to be published).
30. de Gennes, P.G., *Macromolecules* **15**, 492 (1982).
31. Klein, J., *Colloids et Interfaces*, A.M. Cazabat and Md. Veyssie, Eds. (Les Editions de Physique, Paris, 1984).
32. Klein, J., Ingersent, K. and Pincus, P. (to be published).
33. Klein, J. and Pincus, P., *Macromolecules* **15**, 1129 (1982).
34. Ingersent, K., Klein, J. and Pincus, P., *Macromolecules* **19**, 1374 (1986).
35. Rossi, G. and Pincus, P., *Europhysics Letters* (to be published).
36. Almong, Y. and Klein, J., *J. Colloid and Interf. Sci.* **106**, 33 (1985); Klein, J. and Luckham, P., *Nature* **308**, 836 (1984).
37. Israelachvili, J.N., Tirrell, M., Klein, J. and Almog, Y., *Macromolecules* **17**, 204 (1984).
38. Pincus, P.A., Sandroff, C.J. and Witten, Jr., T.A., *J. de Physique* **45**, 725 (1984).
39. Marques, C.M. and Joanny, J.F. (to be published).
40. Wong, K., Cabane, B. and Duplessix, R. (to be published).
41. Cabane, B. and Duplessix, R., *J. Physique* **48**, 651 (1987).
42. des Cloizeaux, J., *J. Physique* **41**, 223 (1980).
43. Hone, D. and Hong Ji, *Macromolecules* (to be published).
44. Langevin, D. and Rondelez, F., *Polymer* **19**, 875 (1978).

RIGOROUS AND EXACT RESULTS FOR A MODEL THREE-COMPONENT SOLUTION

Dale A. Huckaby

Department of Chemistry, Texas Christian University
Fort Worth, Texas 76129

ABSTRACT. A model three-component solution is considered in which the bonds of a lattice are covered by rodlike molecules of types AA, BB and AB. The ends of molecules near a common lattice site interact with energies ϵ_{AA}, ϵ_{BB}, and ϵ_{AB}. In some regions of parameter space, the model is proved to be free of phase transitions. In other regions of parameter space, the model is proved to exhibit phase transitions to ordered low temperature phases. The exact coexistence surface for phase separation into AA-rich and BB-rich phases is calculated for the model on the honeycomb lattice and for a modified version of the model on the square lattice.

1. INTRODUCTION

Twenty years ago Wheeler and Widom[1] introduced a lattice model of a three-component solution in which each bond of a lattice is covered by a rodlike molecule of type AA, BB or AB. The ends of molecules near a common lattice site interact with energy ϵ_{AA} if both ends are of type A, ϵ_{BB} if both ends are of type B, and ϵ_{AB} if one end is of type A and the other end is of type B.

Under the simplifying assumptions $\epsilon_{AB} \to \infty$ and $\epsilon_{AA} = \epsilon_{BB} = 0$, the model can be mapped onto the standard Ising model on the same lattice. The bulk[1] and interfacial properties[2] of this simplified version of the model have been investigated. With the introduction of anisotropic couplings, the model has been used to study the roughening transition[3]. By adding an interaction between neighboring AB molecules, the model can be used to study microemulsions[4]. A six-component version of the model has also been studied[5].

The model with general finite interactions ϵ_{AA}, ϵ_{BB} and ϵ_{AB} was shown to have no phase transitions if $\epsilon_{AB} \le (\epsilon_{AA} + \epsilon_{BB})/2$.[6] For certain ranges of the interaction energies and chemical potentials, the Peierls argument[7] was used to prove the existence of phase separation or of long range order at sufficiently low temperatures for the model on the square and simple cubic lattices[8]. The exact two-phase coexistence surface in temperature-composition space has been obtained for the model with general finite interactions on the honeycomb lattice[9] and for a modified version of the model on the square lattice[10]. The present paper will give an overview of these results for the model with finite interactions.

The Wheeler-Widom model with finite interactions has been generalized by Roble-

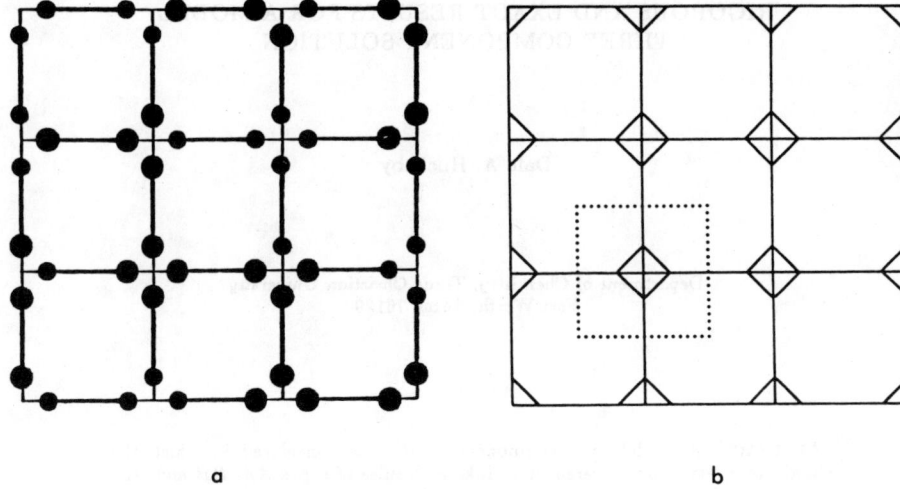

FIGURE 1. (a) Configuration of molecules on $\mathbb{Z}^2$. (b) Line graph Λ_2 associated with the model on $\mathbb{Z}^2$. The square region outlined by a dotted line is referred to in the Peierls argument of section 4.

do[11, 12] to include molecules of types AA, BB and AB. This generalization of the model is equivalent to a spin-3/2 model.[12]

2. ISING REPRESENTATION OF THE MODEL

A complete graph C_v is a graph containing v vertices together with links joining every pair of vertices. A linegraph Λ is a graph which can be covered by a set of complete graphs such that each vertex of Λ is covered by exactly two complete graphs. We shall show that the model with AA, BB and AB molecules with finite interaction is equivalent to a spin-1/2 Ising model on an associated linegraph Λ.[5] A configuration of molecules on $\mathbb{Z}^2$ is illustrated in Figure 1 (a), and the associated linegraph is illustrated in Figure 1 (b)

We can formally write the grand canonical partition function for the model on Λ as

$$\Xi_\Lambda = \sum_\xi e^{-H_\Lambda(\xi)/kT}, \qquad (1)$$

where the Hamiltonian for a configuration ξ is given as

$$H_\Lambda(\xi) = \sum_{X,Y \in \{A,B\}} \left\{ \sum_{(i,j) \subset C_v} \epsilon_{XY} P_i^X P_j^Y - \sum_{(i,j) \subset C_2} \mu_{XY} P_i^X P_j^Y \right\}, \qquad (2)$$

where

$$P_i^X = \begin{cases} 1 & \text{if site } i \text{ is occupied by an } X\text{-type molecular end in } \xi, \\ 0 & \text{otherwise} \end{cases}$$

Since vacant sites are not allowed, the model is considered in the limit in which the chemical potentials μ_{AA}, μ_{BB}, and μ_{AB} all tend to infinity; however, differences such as $\mu_{AB} - \mu_{AA}$ or $\mu_{AB} - \mu_{BB}$ are finite thermodynamic variables which determine the relative concentrations of the three molecular species at equilibrium.[1, 6]

To write the model as a spin-1/2 Ising model, let $S_i = +1$ ($S_i = -1$) if site i is occupied by an A-type (B-type) molecular end. Substituting $P_i^A = (1+S_i)/2$ and $P_i^B = (1-S_i)/2$ into equation (2) yields

$$H_\Lambda(\{S_i\}) = K_I + J_I \sum_{(i,j) \subset C_v} S_i S_j + \mu_I \sum_{(i,j) \subset C_2} S_i S_j - h_I \sum_{i \in \Lambda} S_i, \quad (3)$$

where K_I is a constant, $J_I = (\epsilon_{AA} + \epsilon_{BB} - 2\epsilon_{AB})/4$, $\mu_I = (2\mu_{AB} - \mu_{AA} - \mu_{BB})/4$, and $h_I = (v-1)(\epsilon_{BB} - \epsilon_{AA})/4 - (\mu_{BB} - \mu_{AA})/4$.

3. REGIONS FREE OF PHASE TRANSITIONS

The model can also be written as a lattice gas of A-type molecular ends. Let $t_i = +1$ ($t_i = 0$) indicate that site i is occupied by an A-type (B-type) molecular end. Substitution of $S_i = 2t_i - 1$ into equation (3) gives

$$H_\Lambda(\{t_i\}) = K + 4J_I \sum_{(i,j) \subset C_v} t_i t_j + 4\mu_I \sum_{(i,j) \subset C_2} t_i t_j - h \sum_{i \in \Lambda} t_i, \quad (4)$$

where K is a constant and $h = 2(h_I + (v-1)J_I + \mu_I)$.

Heilmann[13] has shown that the partition function for a repulsive lattice gas on a line graph with equal couplings within each complete graph of the covering set is zero only when $z = e^{h/kT}$ is real and negative. Hence the partition function is never zero for physical values of h so long as $J_I \geq 0$ and $\mu_I \geq 0$.

Ruelle[14] obtained an alternative proof of Heilmann's result which provided the following useful extension. If the couplings on line segments C_2 of the covering set are attractive, the partition function is never zero when $\text{Re}\, z \geq 0$. The partition function is therefore never zero for physical values of h so long as $J_I \geq 0$ and $\mu_i \leq 0$.

The Heilmann and Ruelle results therefore prove that the thermodynamic functions for the Wheeler-Widom model are analytic so long as $J_I \geq 0$. Hence there are no phase transitions if $\epsilon_{AB} \leq (\epsilon_{AA} + \epsilon_{BB})/2$.[6]

In addition, the Lee-Yang circle theorem[15] suffices to show that there are also no phase transitions if $J_I < 0$, $\mu_I < 0$, and $h_I \neq 0$.

4. EXISTENCE OF ORDERED PHASES IN THE MODEL ON Z^D

We shall outline the use of the Peierls argument to prove the existence of ordered phases at low temperatures for certain ranges of the parameters in the model on Z^d, $d = 2$ or 3.[8] (The Peierls argument can also be used to prove the existence of ordered phases in the model on several other lattices.)

As was shown in section 2, the Wheeler-Widom model with finite interaction on Z^d is equivalent to a spin-1/2 Ising model on a line graph Λ_d (see Figure 1). A portion of Λ_d which consists of one C_{2d} graph together with the C_2 graphs which border it will be called a "star". Let H_0 denote the lowest value of the restricted Hamiltonian for a star portion of a configuration. Stars with this value will be called "ground state star". Stars with higher values of the restricted Hamiltonian, H_*, will be called "excited state stars". Let $\alpha_d = \min(H_* - H_0)$.

If $J_I < 0$, $\mu_I < 0$, and $h_I = 0$, the ground state configurations consist of the two configurations containing only AA or only BB molecules. If $J_I < 0$ and $\mu_I > |h_I|$, the ground state configurations consist of the two configurations containing only AB molecules, such that the vertices of each C_{2d} graph are occupied by either all type A or all type B molecular ends. For either of the above ranges of parameters, a consideration of the possible configurations on a star indicates that

$$\alpha_d = \min\{\mu_I - |h_I|, 2(2d-1)|J_I|\} > 0. \tag{5}$$

Since the notion of a contour is central to the Peierls argument, we shall now define what we shall mean by a contour in a configuration. For the two-dimensional lattice Λ_2, the square region r_2 outlined by dots in Figure 1(b) will be said to be associated with the star which has its C_4 portion interior to r_2. Analogously in three dimensions, a cubic region r_3 will be associated with each star in Λ_3. If r_d is associated with an excited state star, then r_d is said to be a contour segment. Two contour segments will be said to be connected in the two-dimensional case if they share a common vertex and in the three-dimensional case if they share a common edge.

If N_δ is the number of bonds of Z^d which contain molecules which belong to an ordered structure δ on a lattice containing $|\Lambda_d|/2$ such bonds, then as discussed by Dobrushin[16] and Griffiths,[17] there is an ordered phase in the thermodynamic limit if $\langle N_\delta \rangle/(|\Lambda_d|/2) > 1/2$, where there are two symmetrically related ordered structures, and where the thermal average is taken only over configurations in which the outer boundary of the lattice contains the ordered structure δ. Convergence proofs of series expansions suffice to show that the model is disordered at high temperatures, thereby proving the existence of an order-disorder transition in the model.

If the stars on the boundary are occupied by one of the ground state structures δ discussed above, then all bonds of Z^d containing molecules not belonging to δ are enclosed by (or embedded within) an outer contour γ. As a consequence,

$$|\Lambda_d|/2 - \langle N_\delta \rangle < \sum_{|\gamma|} m_{|\gamma|} n_{|\gamma|} P_\gamma, \tag{6}$$

where $m_{|\gamma|}$ is the maximum number of molecules enclosed by (or embedded in) a contour of length $|\gamma|$, $n_{|\gamma|}$ is the number of contours containing $|\gamma|$ segments, and P_γ is an upper bound to the probability a given contour of length $|\gamma|$ occurs in a configuration.

The number of possible closed contours containing $|\gamma|$ segments is bounded as $n_{|\gamma|} < a_d c_d^{|\gamma|} |\Lambda|/|\gamma|$, where a_d is a constant, c_d is the maximum number of ways to continue a contour, and $|\Lambda|$ is proportional to the number of places at which a contour can be begun. The division by $|\gamma|$ results since the choice of the first segment is arbitrary. The number of molecules contained in a contour is bounded above by the number of molecules within a d-dimensional isoperimetric spherical region. Hence $m_{|\gamma|} < b_d |\gamma|^{d/(d-1)}$, where b_d is a constant.

With each configuration ξ which contains an outer contour γ we can associate (this is a 1-1 correspondence) a configuration ξ^* which does not contain γ and which has the property $H_{\Lambda_d}(\xi) - H_{\Lambda_d}(\xi^*) \geq \alpha_d|\gamma|$.[5] Hence $P_\gamma < e^{-\alpha_d|\gamma|/kT}$.

Equation (6) then ensures that $\langle N_\delta \rangle/(|\Lambda_d|/2) > 1/2$ at sufficiently low temperatures. Therefore, there are AA-rich and BB-rich phases in the model at sufficiently low temperature if $J_I < 0$, $\mu_I < 0$, and $h_I = 0$. In addition, if $J_I < 0$ and $\mu_I > |h_I|$, there is an ordered AB phase in the model at sufficiently low temperatures.[8]

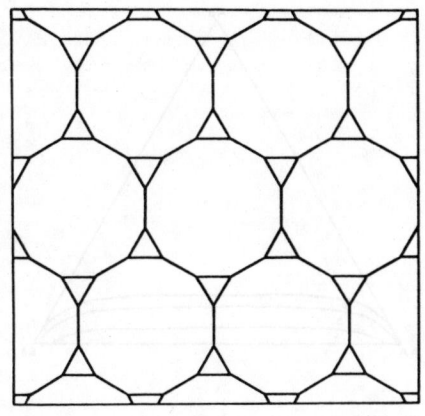 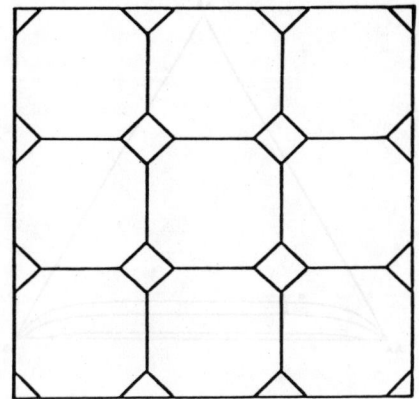

FIGURE 2. (a) The 3–12 lattice. (b) The 4–8 lattice.

5. EXACT TWO-PHASE COEXISTENCE SURFACES

The mole fractions of AA, BB and AB molecules in the model can be calculated from the relationships[9, 10]

$$X_{AA} + X_{BB} + X_{AB} = 1$$
$$|X_{AA} - X_{BB}| = I \qquad (7)$$
$$X_{AB} = (1 - \sigma)/2$$

Here I is the magnetization of the associated Ising model on Λ, and

$$\sigma = \langle S_i S_j \rangle_{i,j \subset C_2}.$$

(Note that $S_i S_j = 1$ if an AA or BB molecule covers C_2, and $S_i S_j = -1$ if an AB molecule covers C_2.)

As indicated in section 4, phase separation into AA-rich and BB-rich phases occurs at sufficiently low temperature if $h_I = 0$, $J_I < 0$, and $\mu_I < 0$. The corresponding two-phase coexistence surface in temperature-composition space has been calculated exactly for the model on the honeycomb lattice[9] and for a modified version of the model on the square lattice.[10]

As demonstrated in section 2, the molecular model on the honeycomb lattice is equivalent to an Ising model on the 3–12 lattice, which is illustrated in Figure 2 (a). A modified version of the model on the square lattice in which only first-neighbor molecular ends of molecules near a common site interact is equivalent to an Ising model on the 4–8 lattice, which is illustrated in Figure 2 (b). (Note that the 4–8 lattice, corresponding to the modified version of the model, is not a line graph.)

Using the star-triangle and multiple decoration-iteration transformations, the partition

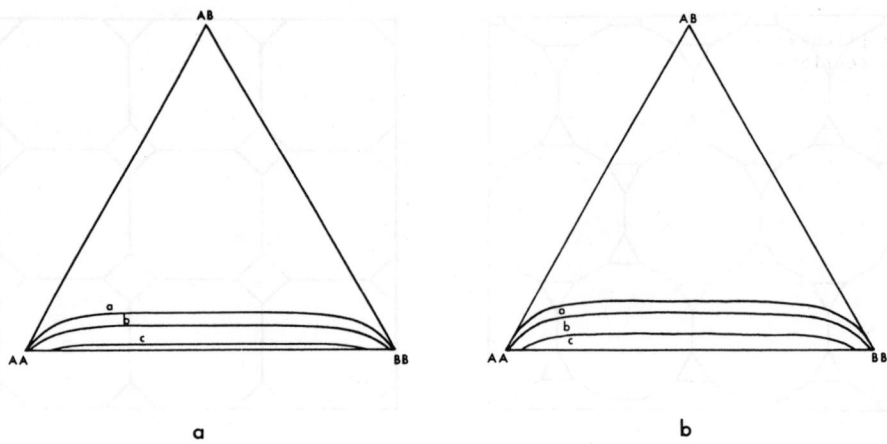

FIGURE 3. Isothermal coexistence curves at the temperatures $T'_a \to 0$, $T'_b = 1.5$, $T'_c = 2.0$ for (a) the model on the honeycomb lattice and (b) the modified version of the model on the square lattice.

function of an Ising model on the 3–12 lattice can be related to that of an Ising model on the honeycomb lattice.[9,18,19] This result enabled us to calculate exact expressions for I and σ when $h_I = 0$.[9,19]

Using the partition function for a zero-field $(h_I = 0)$ Ising model on the 4–8 lattice, we were able to calculate an exact expression for σ.[10] Other workers have given an expression for the spontaneous magnetization of an Ising model on the 4–8 lattice.[20] Since their expression for I reproduces the exact low-temperature series expansion for I up to at least twelfth order, and since it gives exact results for several limiting cases, they have conjectured that their expression is exact.[20,21]

Exact two-phase coexistence curves at three different temperatures are illustrated in Figure 3 (a) for the model on the honeycomb lattice and in Figure 3 (b) for the modified model on the square lattice. Letting $T' = kT/|J_I|$, the coexistence surface for the model on the honeycomb lattice has the limiting values $\max T' = 4/\ln(3 + 2\sqrt{3}) = 2.143\ldots$ and $\max X_{AB} = (9 - 4\sqrt{3})/18 = 0.1151\ldots$.[9] The corresponding limiting values for the coexistence surface of the modified model on the square lattice are $\max T' = 2/\ln(\sqrt{2} + 1) = 2.269\ldots$ and $\max X_{AB} = (2 - \sqrt{2})/4 = 0.1464\ldots$.[10]

At present we are studying the model on an infinite three-coordinate Bethe lattice.[22] We hope to publish the results of this study in the near future.

ACKNOWLEDGMENT

Support from The Robert A. Welch Foundation Grant P-446 is gratefully acknowledged.

REFERENCES

1. Wheeler, J.C. and Widom, B., *J. Am. Chem. Soc.* **90**, 3064 (1968).
2. Widom, B., *J. Phys. Chem.* **88**, 6508 (1984).
3. Robledo, A., Varea, C. and Martina, E., *J. Physique Lett.* **46**, 967 (1985).

4. Widom, B., *J. Chem. Phys.* **84**, 6943 (1986).
5. Varea, C. and Robledo, A., *Phys. Rev. A***33**, 2760 (1986).
6. Huckaby, D.A. and Kowalski, J.M., *Phys. Rev. A***30**, 2121 (1984).
7. Peierls, R., *Proc. Camb. Phil. Soc.* **32**, 477 (1936).
8. Huckaby, D.A., Kowalski, J.M. and Shinmi, M., *J. Phys. A.* **18**, 3585 (1985).
9. Huckaby, D.A. and Shinmi, M., *J. Stat. Phys.* **45**, 135 (1986).
10. Shinmi, M. and Huckaby, D.A., *J. Phys. A* **20**, L465 (1987).
11. Robledo, A., *Europhys. Lett.* **1**, 303 (1986).
12. Robledo, A., *Phys. Rev. A* **36**, 4067 (1987).
13. Heilmann, O.J., *Stud. Appl. Math.* **50**, 385 (1971).
14. Ruelle, D., *Commun. Math. Phys.* **31**, 265 (1973).
15. Lee, T.D. and Yang, C.N., *Phys. Rev.* **87**, 410 (1952).
16. Dobrushin, R.L., *Funct. Anal. Appl.* **2**, 302 (1968).
17. Griffiths, R.B., *Phase Transitions and Critical Phenomena*, Vol. I, C. Domb and M. Green, Eds. (Academic Press, New York, 1972).
18. Syozi, I., *Phase Transitions and Critical Phenomena*, Vol. I, C. Domb and M. Green, Eds. (Academic Press, New York, 1972).
19. Huckaby, D.A., *J. Phys. C* **19**, 5477 (1986).
20. Lin, K.Y., Kao, C.H. and Chen, T.L., *Phys. Lett. A* **121**, 443 (1987).
21. Lin, K.Y., *J. Stat. Phys.* **49**, 269 (1987).
22. Baxter, R.J., *Exactly Solved Models in Statistical Mechanics* (Academic Press, New York, 1982).

EFFECT OF THE COUNTERIONS ON THE ELECTROKINETIC PROPERTIES AND ION ADSORPTION IN CHARGED MICROPOROUS MEDIA

Wilmer Olivares, Marlene Huerta, Pedro Colmenares and Juan C. Villegas

*Grupo de Química Teórica, Departamento de Química,
Facultad de Ciencias, Universidad de Los Andes,
Mérida – Venezuela*

ABSTRACT. We present the numerical solution of the non linear Poisson-Boltzmann equation for the ionic distribution inside a charged cylindrical micropore. We discuss the importance of the presence of the dissociated counterions on the potential profiles and hence on electrokinetical properties of the system, obtaining a good agreement with the existing experimental data. The adsorption of the counterions at the charged surface is considered through a dissociation constant and the zeta potential is calculated in a self consistent manner for low surface charge densities.

1. INTRODUCTION

An understanding of electrokinetic flow through narrow microcapillaries is of considerable importance in many fields of science, particularly in chemistry, medicine and engineering.[1] For instance, a network of microcapillaries is known to represent well many porous media as underground oil reservoir rocks, sand beds, molecular sieves, organic tissues, blood microcapillaries, catalysts as zeolites, etc.

The simplest model of these systems is obtained by considering an infinitely long charged cylinder inmersed in an electrolyte solution which invades its interior. The charge is assumed to be uniformly smeared at the inner surface of the cylinder. The application of potential and pressure gradients along such a device gives a good description of the experimentally observed streaming potential, electro-osmosis and electroviscous retardation effects.[2]

Most of the previous work in this field considers that the surface charge arises from the total dissociation of ionizable groups located at the inner surface of the pore or from the adsorption of charged species from the solution. The pore radius R or the added salt concentration, and therefore the Debye-Hückel screening parameter κR, are usually considered to be large enough that the counterions coming into solution as a consecuence of the dissociation have a negligible effect on the properties of the system. However, we have found that physically significant values of the charge σ or surface potential Ψ_R, the counterions have a very important effect on the ionic structure within the micropores and hence on all the thermodynamic and electrokinetical properties of the ionic solution.[3,4]

In this work we have solved numerically the appropriate non linear Poisson-Boltzmann equation previously discussed by us and referred to as the $\kappa\gamma$-equation. This exact solution allows the study of the potential profiles and of the double layer structure inside a charged micropore for large surface charge densities and small pore radii of practical interest. We then analyze the effect of this structure on electrokinetical properties such as the electroosmotic and electroviscosity coefficients which correct the zeta potential and the solution viscosity respectively. Finally, we study a situation where the counterions do not dissociate completely but have a finite equilibrium constant associated to the dissociation process and then calculate the zeta potential in a self consistent manner.

2. GENERAL THEORY

The validity and limitations of the non linear Poisson-Boltzmann equation (PB) are well stablished for electrolyte solutions and double layer problems.[5-8] For such inhomogeneous ionic systems, the PB equation is obtained when the ions are considered as point charges and the ion-ion short range correlations are neglected.[5-8] The potential of mean force acting on a given ion under an external electrical field is then approximated by the mean electrostatic potential $w_i e \Psi$. The resulting equation contains most of the physics of the system and is useful for obtaining a first insight into complex systems:

$$\nabla^2 \Psi(r) = -\frac{4\pi}{\epsilon} \sum_i w_i e \rho_i e^{-\beta w_i e \Psi(r)} \tag{1}$$

Here $w_i e$ and ρ_i are the charge and particle density of ionic species i. ϵ is the dielectric constant of the medium, assumed to be uniform, and $\beta = 1/kT$ is the reduced inverse absolute temperature.

For the planar and cylindrical electrical Double Layers it has been found that for low surface charge densities, low concentrations and low values of the parameter $w^2 e^2 / \epsilon k T$ the PB equation fares very well with the results of more rigorous theories as the HNC[7] and WLMB[8] approximations. This equation has been used extensively[1-4] to study model cylindrical polyelectrolyte solutions (outer problem), and to describe the electrokinetical properties of cylindrical micropores (inner problem). As mentioned before, we are interested in the inner problem represented in Figure 1. For this system the PB equation becomes

$$\nabla^2 \Psi(r) = -\frac{4\pi}{\epsilon} \left[\sum_i^* w_i e \rho_i e^{-\beta w_i e \Psi} + w_c e \rho_c e^{-\beta w_c e \Psi} \right] \tag{2}$$

where $\sum^*$ denotes the sum over all the ionic species of the added electrolyte and the second term is the contribution to the charge density of the counterions in solution that balance the surface charge at the cylinder's wall, assumed to arise from the dissociation of some given ionizable groups. If the equivalent smeared surface charge density is σ, we should have $\rho_c = 2\sigma/R$, by electroneutrality.

For a symmetrical $w - w$ electrolyte and counterions with charge $w_c = w$, equation (1) becomes, in reduced units

$$\nabla^2 \phi = \kappa^2 \sinh \phi + \gamma^2 e^\phi \tag{3}$$

where $\kappa^2 = 8\pi\beta w^2 e^2 \rho / \epsilon$ and $\gamma^2 = 8\pi\beta w e \sigma / R$ represent the reduced concentration and surface charge respectively. $\phi(r)$ is the reduced mean electrostatic potential, $\phi = \beta w e \Psi$. Here $\rho = \rho_+ = \rho_-$ is the electrolyte density. This equation is subjected to the boundary conditions

$$\left(\frac{d\phi}{dr} \right)_{r=R} = \frac{\gamma^2 R}{2} \tag{4}$$

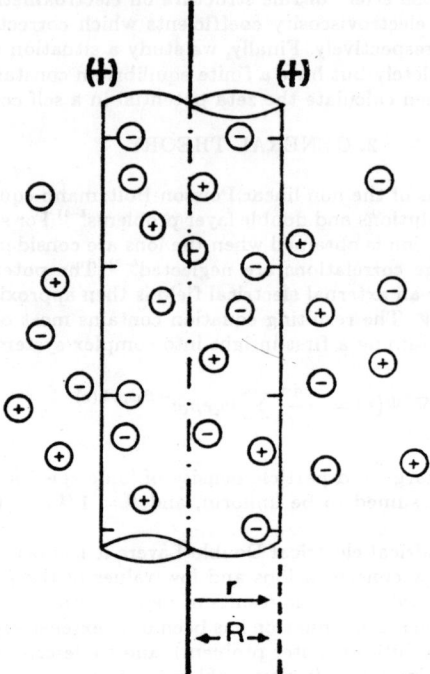

FIGURE 1. Restricted primitive model of a charged micropore in equilibrium with an electrolyte solution.

and $(d\phi/dr) = 0$ at $r = 0$. For high electrolyte concentrations and/or low surface charges $\kappa/\gamma \gg 1$ and equation (3), referred to as the $\kappa\gamma$-equation, reduces to the so called κ-equation commonly used in the literature[1-4]

$$\nabla^2 \phi = \kappa^2 \sinh \phi \qquad (5)$$

The main argument here is that, for most practical values of κR and γR, the counterion term appearing both in the differential equation and in the boundary condition is very important and should not be neglected.

In a previous work,[3] we solved equations (3) and (4) by the variational procedure of Arthurs and Robinson.[2] The interesting aspects of such solution are that the functional which we have to minimize is identical to the Helmholtz free energy of the system and that a one parameter trial function obtained from the linearized equation is enough to obtain reasonably accurately results.

It is interesting to point out that a simple function transformation of equation (3) gives

$$\nabla^2 \psi = \alpha^2 \sinh \psi \qquad (6)$$

where

$$\psi = \phi + \ln t \tag{7}$$
$$\alpha^2 = \kappa^2 t \tag{8}$$

and

$$t = (1 + 2\gamma^2/\kappa^2)^{1/2} \tag{9}$$

Since equation (6) has the same functional form of equation (5), we have been able to use the solutions given by Levine et al.[1] and by Oldham et al.,[10] with the appropriate parameters to obtain rapid approximate solutions of equation (3). However, since the accuracy of such solutions depends on the range of the values of κR (or αR) and γR, an exact numerical solution is still necessary.

The PB equation was then solved using the routines **PASVA3**[11] and **CLOSYS**[12] which are based on a method of deferred corrections with automatic step and gridding control.

As we will see next, all the properties of interest depend on simple volume integrations of functions depending on the potential profile $\phi(r)$, which were carried out using a Gaussian Cuadrature. For consistency, the ion-wall activity coefficients were calculated using our previous result[3] obtained from the Free energy associated to equation (3)

$$\ln \gamma_\pm = \frac{1}{V} \int_v (\cosh \phi - 1) \, dr \tag{10}$$

Finally, we have considered the situation where the counterions A^+ are not completely dissociated from the wall, by allowing a dissociation equilibrium

$$S - A \Longleftrightarrow S^- + A^+ \tag{11}$$

to occur, with a dissociation constant K_A. The charge is then found as $\sigma = e\nu_s \delta$, where ν_s is density of ionizable sites and δ is the degree of dissociation of the sites, obtained from the equilibrium as

$$\delta = K_A/(K_A + \rho\gamma_+) \tag{12}$$

where ρ is the salt concentration and γ_+ is the counterion activity coefficient. The potential is then obtained self consistently[13] by solving simultaneously Eqs (3), (10) and (12) with the boundary condition

$$\left(\frac{d\phi}{dr}\right)_{r=R} = \frac{4\phi}{\epsilon}\nu_s\delta\beta w^2 e^2 \tag{13}$$

In this work we present only the results for the Rice and Whitehead[14] limit, obtained from the solution of the linearization of equation (3), valid for low surface charges but having an analytical solution

$$\phi = C_1 I_0(br) + (\gamma^2/b)^2 \tag{14}$$

where I_0 is the modified Bessel function of zero order and $b^2 = \kappa^2 + \gamma^2$. The integration constant C_1 and the degree of dissociation δ are obtained using equations (10)–(13).

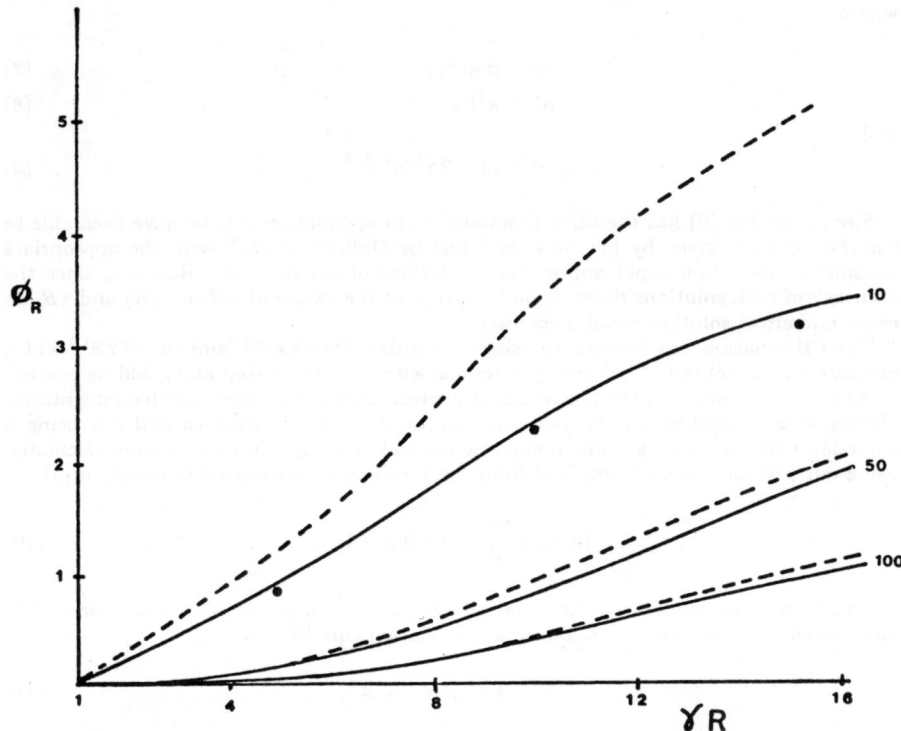

FIGURE 2. Reduced contact potential as a function of the reduced surface charge for several values of κR. —— $\kappa\gamma$-equation, --- κ-equation solved variationally[?] (•) numerical solution of the $\kappa\gamma$-equation.

3. RESULTS AND DISCUSSION

In Figure 2 we show the potential at the surface as a function of the charge parameter γR for several values of the added salt concentration parameter κR. We can see that the κ-equation, obtained by neglecting the counterions term in equation (3), predicts potentials that are too high as compared to those predicted by the $\kappa\gamma$-equation. The potential profiles $\phi(r)$ versus r presented in Figure 3 show that, for low charges, while the κ-equation predicts small changes as one moves from the wall to the midpoint of the pore, the $\kappa\gamma$-equation predicts a drastic change in the potential and a charge inversion is suggested by the change in sign of the potential. The tendency of the midpotential to become the negative of the surface potential as the charge increases, agrees with the results obtained by solving analytically the equation resulting from neglecting the first term in equation (3) (γ-equation) which corresponds to a situation of having the counterions alone.

This charge inversion can be interpreted as a packing of the ions of opposite (−) sign about the (+) surface presenting an effective negative charge to the electrolyte solution at the midpoint of the micropore, as represented in Figure 2. As a consecuence, the system

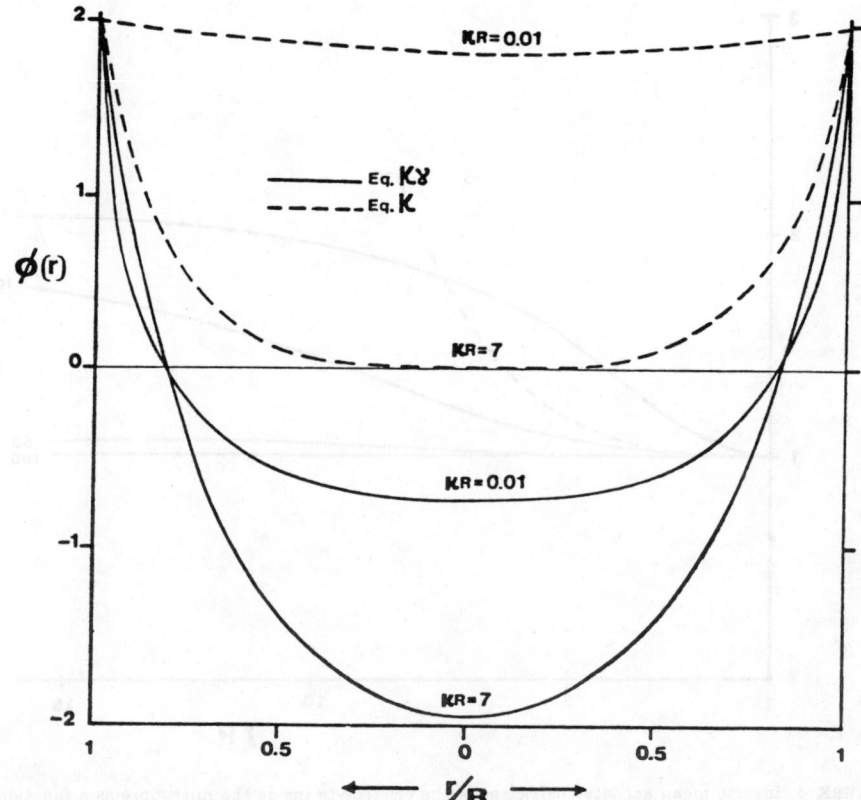

FIGURE 3. Potential profiles $\phi(r)$ versus r/R for $\phi_R = 2$. (——) corresponds to equation (3) and (- - -) to equation (5).

is inhomogeneous at all points inside the micropore and there is not a "bulk solution point", as is erroneously predicted by the κ-equation, for large γR.

So, while the simple variational solution gave surface potentials that fared very well with the numerical values for $\phi_R < 4$ and $\gamma R < 10$, as shown in Figure 2, the numerical solution allowed us to study the details of the fluid structure inside the micropore.

The effect of the observed behavior on the wall-ion activity coefficient is shown in Figure 4. The κ-equation predicts excedingly large values while all our $\kappa\gamma$-equation predictions have physically reasonable values.[5]

An important effect observed in this kind of system is the so called ion exclusion, measured by the distribution coefficient D, which is given by the ratio of the coions concentrations in, ρ_+, and out, ρ'_+, of the micropore

$$D = \rho_+/\rho'_+ = \gamma_\pm/t\gamma'_\pm \qquad (15)$$

where $\gamma_\pm$ and $\gamma'_\pm$ are the mean activity coefficients of the solutions in and out of the micropore and t is given by equation (8). Figure 5 shows D versus κR for several values of R. We can see that the coions are strongly excluded, namely $D \ll 1$, when γR becomes

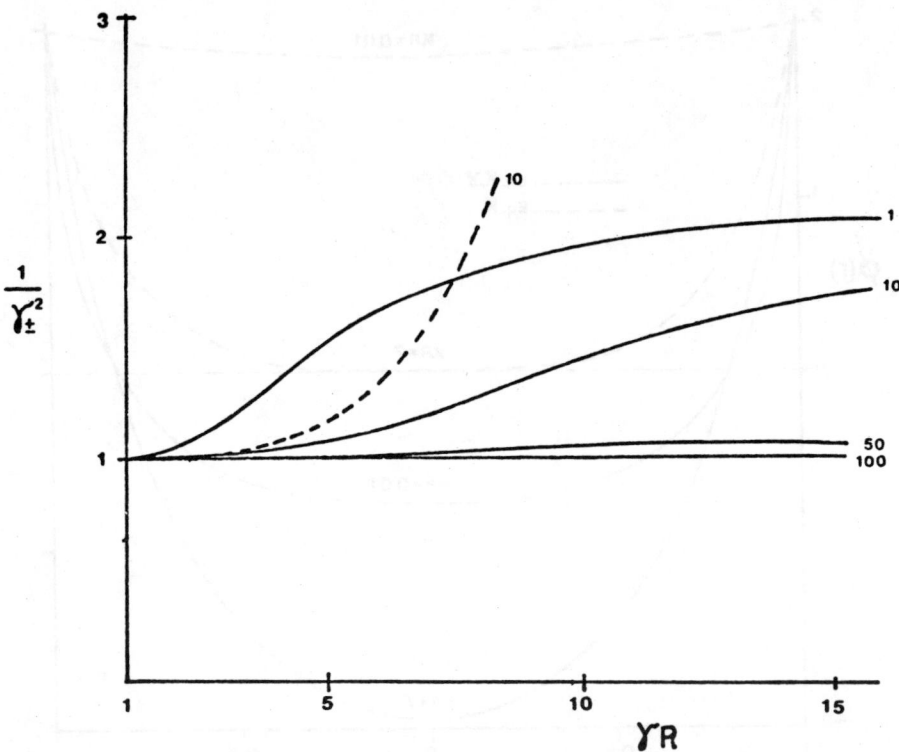

FIGURE 4. Inverse mean activity coefficient of the electrolyte inside the micropore as a function of the reduced charge γR for several values of κR. (- - -) is the result from equation (5).

large or κR is small (high surface charge or small capillaries). Neglecting the counterion effect would give unphysical values ($D > 1$) for large γR.[4]

When an electrical field E is applied along a micropore, one obtains a net volumetric flow given by

$$V = \dot{V}_\infty (1 - G) \tag{16}$$

where $\dot{V}_\infty$ is the Smoluchowsky limit,[1] obtained for very large micropores, and G is given by

$$G = \frac{2}{\phi_R R^2} \int_0^R r\phi(r)\,dr \tag{17}$$

In Figure 6 we show $1 - G$ as a function of κR for two values of the zeta potential. As can be seen a completely different behavior is obtained with equation (3) as compared to equation (5), as a consecuence of the very different potential profiles.

In electroosmosis experiments the zeta potential ϕ_R is corrected by a factor F which

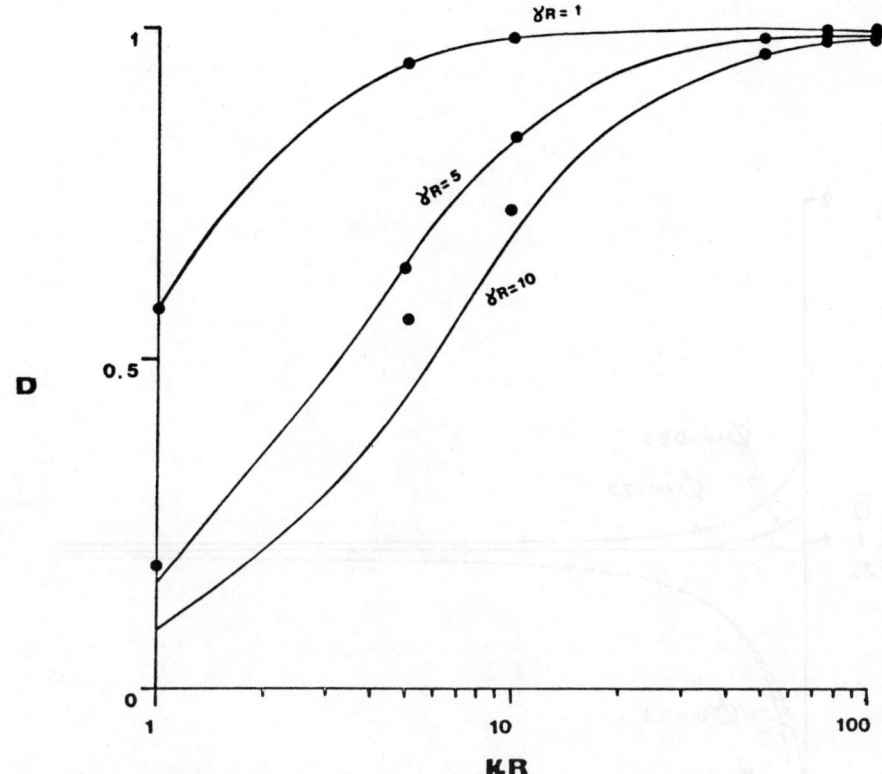

FIGURE 5. Distribution coefficient versus κR for several reduced charges. A comparison is made between the variational (——) and numerical (•) solutions.

can be measured experimentally.[2] We have that

$$F = (1 - G)/A \tag{18}$$

where

$$A = \cosh \phi_R + \frac{\gamma^2}{\kappa^2}\left(e^{\phi R} - \frac{\gamma^2 R^2}{\delta}\right) + \frac{2\beta^*}{\kappa^2 R^2} \int_0^R r(d\phi/dr)^2 \, dr \tag{19}$$

where β^* is a constant that depends on the electrolyte solution properties and is estimated as 0.25.

In Figure 7 we compare the theoretical predictions with the F coefficient measured by Bröz and Epstein[2] in glass borosilicates microcapillaries ($\phi_R = 68$) and in nucleopore polycarbonate membrane filter ($\phi_R = 2.2$). As can be seen a good agreement is found within the experimental error.

In Figure 8 we show the effect of a finite dissociation constant K_A on the self consistent surface potential ϕ_R, for two densities of ionizable groups, as a function of the electrolyte

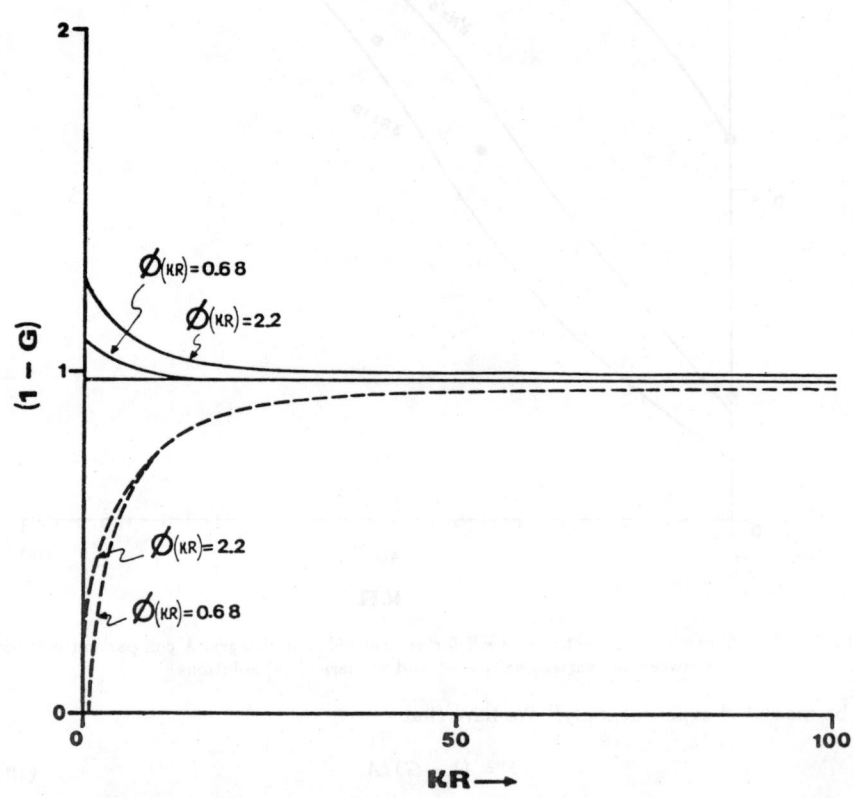

FIGURE 6. Volumetric flow coefficient $1 - G$ versus κR for two values of the contact potential. (———) corresponds to equation (3) and (- - -) to equation (5).

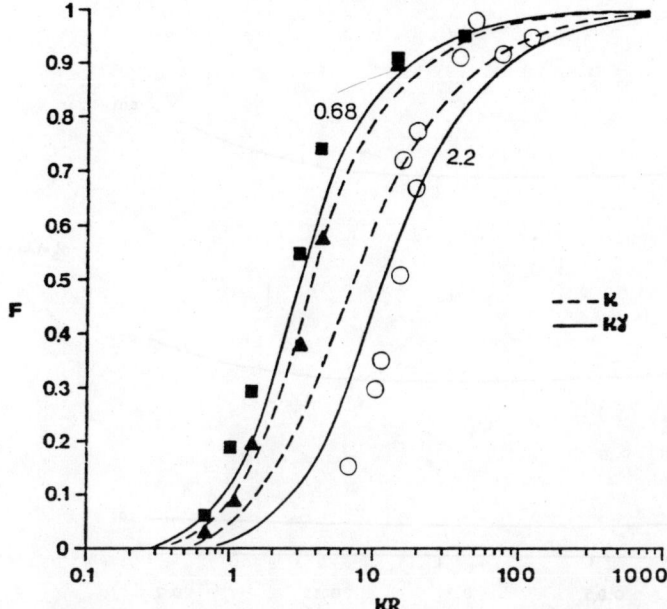

FIGURE 7. Electroosmosis factor F versus κR for two contact potentials. ($\square$) and ($\triangle$) represent the experimental data for $\phi_R = 0.68$ and ($\bigcirc$) corresponds to the data for $\phi_R = 2.2$.[2]

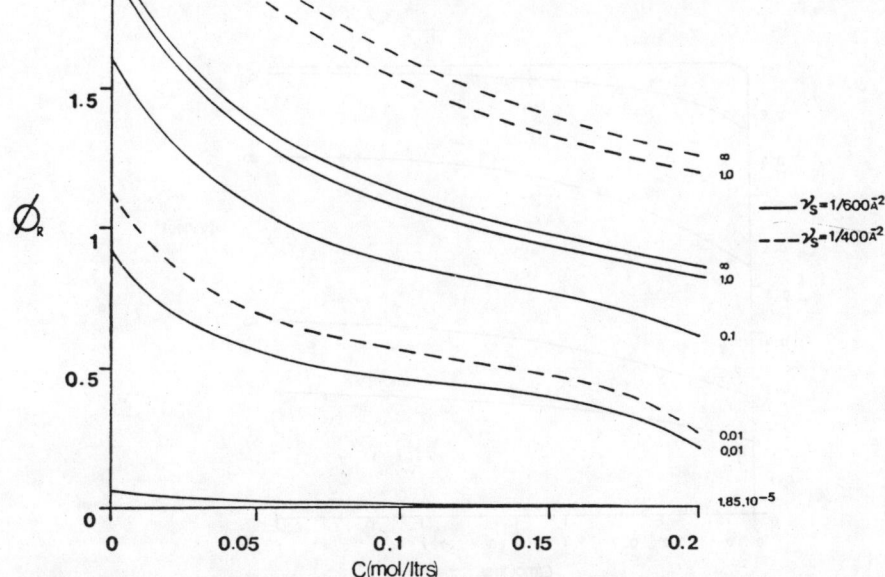

FIGURE 8. Contact potential versus concentration for several values of the equilibrium constant K_A and two densities of ionizable groups.

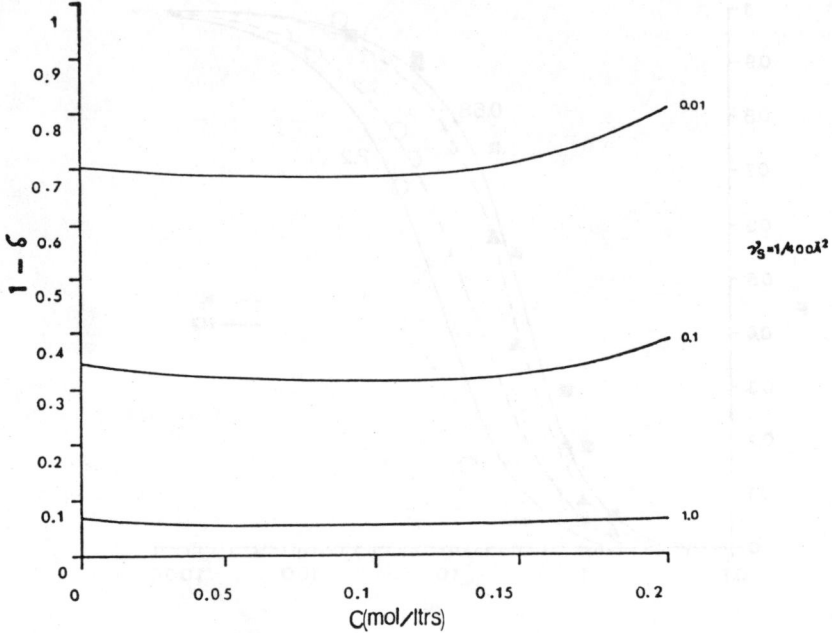

FIGURE 9. Degree of association as a function of salt concentration for several values of K_A. The added electrolyte is assumed not to adsorb to the surface.

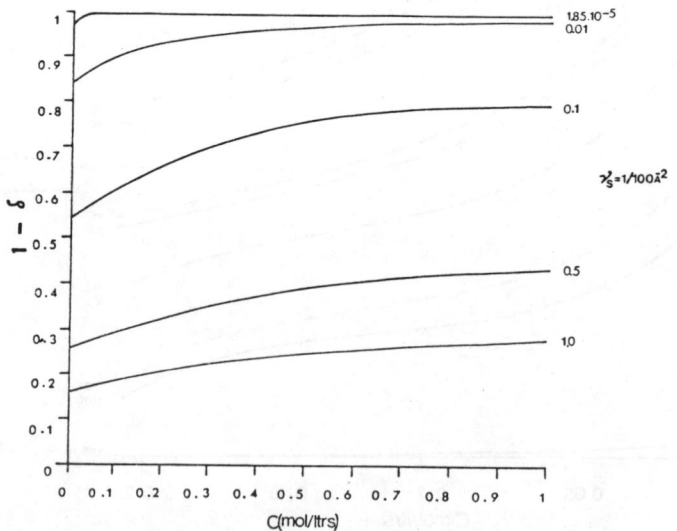

FIGURE 10. Degree of association as a function of salt concentration for several values of K_A. The counterions dissociated and the added salt cations are identical.

concentration inside a micropore of 100 Å in radius. The strong decrease of the surface potential as the dissociation constant decreases is an interesting effect that helps in the understanding of the ion exchange in porous media. When two ionic species compete for the same sites at the surface, the selectivity will depend mainly on the dissociation constants.

Here we have only considered two limiting situations for a negatively charged surface. In the first case the added salt contains a cation whose dissociation constant is infinitely large compared to that of the dissociated counterions. The second case considers that the cations of the salt are identical to the dissociated counterions. In the first case, shown in Figure 9 we simply observe a salinity effect on the association isotherm with an essentially constant degree of association for small concentrations. In the second case we see a common ion effect on the adsorption isotherm. In Figure 10, the degree of association $1 - \delta$ is plotted as a function of the salt concentration for several values of the dissociation constant, showing the expoected Langmuir isotherms type fo behavior.

4. CONCLUSIONS

We have solved the PB equation numerically to obtain the potential profiles $\Psi(r)$ versus r for the inner problem of an electrolyte solution in a charged cylindrical micropore. A charge inversion was found when the dissociated counterions in solution where explicitly included in the differential equation. This structure of the double layer inside a charged micropore leads to physically significant values for the zeta potentials, the activity coefficients and the distribution coefficient for high surface charge densities and/or low screening parameter κR, contrary to the predictions obtained by neglecting the counterion effect.

The predicted electrokinetical properties agree reasonably well with the measured data within the experimental error.

The self consistent calculation of the surface charge for finite dissociation constants of the ionizable groups allowed the calculation of physically meaningful adsorption type isotherms that will prove to be useful in the interpretation of the ion exchange isotherms in porous media. The consideration of a second type of ion with a finite dissociation constant is an immediate extension that will be the subject of future reports.

ACKNOWLEDGMENTS

This work was supported by the Consejo de Desarrollo de la Universidad de Los Andes, Grant CDCH-C235. We acknowledge many helpful discussions with D. Morales.

REFERENCES

1. Levine, S. and Neale, G., *J. Colloid. Interf. Sci.* **47**, 520 (1974); Levine, S., Marriott, J.R., Neale, G. and Epstein, N., *J. Colloid. Interf. Sci.* **52**, 136 (1975).
2. Broz, Z. and Epstein, N., *J. Colloid. Interf. Sci.* **56**, 605 (1976).
3. Huerta, M. and Olivares, W., *J. Phys. Chem.* **91**, 2975 (1987).
4. Dresner, L., *J. Phys. Chem.* **67**, 2333 (1963).
5. Fixman, M., *J. Chem. Phys.* **70** 4995 (1979).
6. Medina-Noyola, M. and Keiser, J., *Physica* **107A**, 438 (1981).
7. Lozada-Cassou, M., *J. Phys. Chem.* **87**, 3729 (1984); González-Tovar, Lozada-Cassou, M. and Henderson, D., *J. Chem. Phys.* **83**, 361 (1985).
8. Colmenares, P. and Olivares, W., *J. Chem. Phys.* (in press).
9. Arthurs, A. and Robinson, P.A., *Proc. Cambridge Phil. Soc.* **65**, 535 (1969).
10. Oldham, I.B., Young, F.J. and Osterle, J.F., *J. Colloid. Interf. Sci.* **18**, 328 (1963).

11. Lentini, M. and Pereira, V., *Siam J. Num. Anal.* **14**, 91 (1977).
12. Russel, R.D. and Christiansen, J., *Siam J. Num. Anal.* **7**, 59 (1978).
13. Brenner, S. and McQuarrie, D.A., *J. Theor. Biol.* **39**, 343 (1973).
14. Rice, C.L. and Whitehead, R., *J. Phys. Chem.* **69**, 4017 (1965).

ANISOTROPIC STRUCTURE OF A SIMPLE LIQUID[*]

H.J.M. Hanley

Thermophysics Division, National Bureau of Standards, Boulder, CO 80303

ABSTRACT. Non-Newtonian behavior in a sheared simple fluid is reviewed briefly. We point out that such behavior is more often associated with fluids of very complex structure, such as polymers. The concept of the Maxwell relaxation time is introduced and it is shown that the product of this time and a shear rate is a key parameter that indicates if a fluid will display non-Newtonian characteristics, regardless of molecular structure. We discuss the distorted structure factor of a dense two-dimensional soft disk liquid undergoing Couette flow. The structure factor is determined experimentally by light scattering from the output of a nonequilibrium molecular dynamic simulation. The distorted pair correlation function is also discussed.

key words: Non-Newtonian fluid; structure factor; pair correlation function; nonequilibrium molecular dynamics; two-dimensional soft disks.

1. INTRODUCTION

In this paper we review one of the more significant results of nonequilibrium molecular dynamics; namely, that simple fluids under shear can display non-Newtonian features that one sees with polymers, paint, or bread dough, or other very complex liquids. The paper is organized in two parts. We first discuss very briefly what is generally meant by non-Newtonian behavior in a sheared liquid, and present typical non-Newtonian results from model simple liquids obtained from nonequilibrium molecular dynamics (NEMD). Second, we discuss the shear distorted intermolecular microstructure of a model liquid observed by the structure factor, $S(\mathbf{k}, \gamma)$, where $\mathbf{k}$ is the wave vector and γ the shear rate: a proper explanation of phenomena in a sheared fluid must involve this quantity.

2. NON-NEWTONIAN BEHAVIOR

Simple and complex liquids can behave quite differently under normal laboratory conditions. The Weissenberg effect is a well-known example. Figure 1 demonstrates what happens when a simple (Newtonian), and a complex (non-Newtonian) fluid are stirred by a rod. For the moment "simple and Newtonian" and "complex and non-Newtonian" are used interchangeably. The surface of the simple fluid is depressed largely because of the centrifugal force, whereas the complex fluid overcomes the force and climbs the rod.

[*]Publication of the National Bureau of Standards: not subject to copyright.

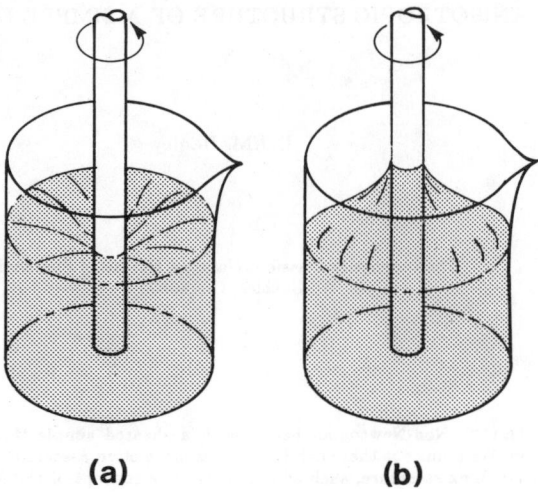

FIGURE 1. Sketch of the Weissenberg effect. In (a) a Newtonian liquid is stirred and the meniscus is depressed at the rod, whereas in (b) the non-Newtonian liquid climbs the rod. Sketched from figures presented in reference 1.

Figure 2 shows a siphon. The siphoning of a simple liquid will stop if the tube leaves the surface of the reservoir liquid, but one can siphon a complex liquid even if there is a gap between the reservoir liquid and the tube.

The apparent abnormal and exotic behavior of the complex liquid in these, and other examples[1] can be explained if:

1. The viscosity of the liquid, η_+, depends on the shear rate, *i.e.*,

$$p_{xy} = -\eta_+(\gamma)\gamma \tag{1}$$

where p_{xy} is the xy stress component of the pressure tensor $\underset{\approx}{\mathbf{p}}$.

2. That the viscosity is frequency dependent and the fluid is viscoelastic.
3. Normal pressure differences exist on the sheared fluid, *i.e.*,

$$p_{xx} \neq p_{yy} \neq p_{zz} \tag{2}$$

(and that these normal components of $\underset{\approx}{\mathbf{p}}$ are shear rate dependent).

Equations (1)–(3) are, of course, the basis for the discipline of rheology and there has been an enormous body of work on their applications to the flow of polymers and complex liquids. What is recent is that we have acquired *quantitative* data from computer simulation that simple liquids (liquids with spherical molecules) also have shear rate dependent viscosities, are viscoelastic and display normal pressure differences.

Examples of nonequilibrium molecular dynamics results are given in Figures 3, 4 and 5. The data were extracted from simulations of a soft sphere, inverse-twelve, liquid undergoing Couette flow.[2] The results are typical and are qualitatively independent of the nature of the potential and the computer algorithm. Figure 3 depicts the shear viscosity

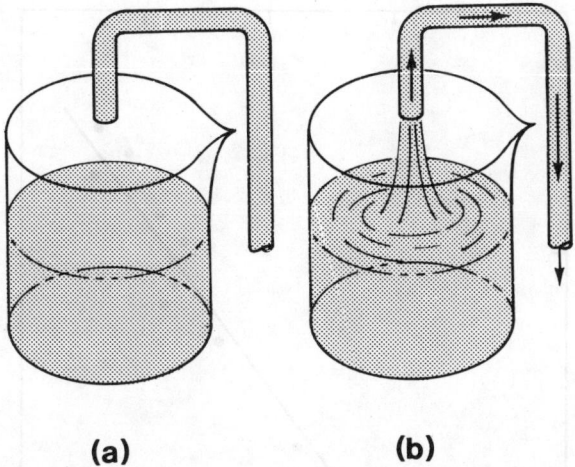

FIGURE 2. Sketch of the difference between a simple (a) and complex (b) liquid when siphoned. In (b), siphoning will continue even if the tube is raised out of the reservoir liquid.

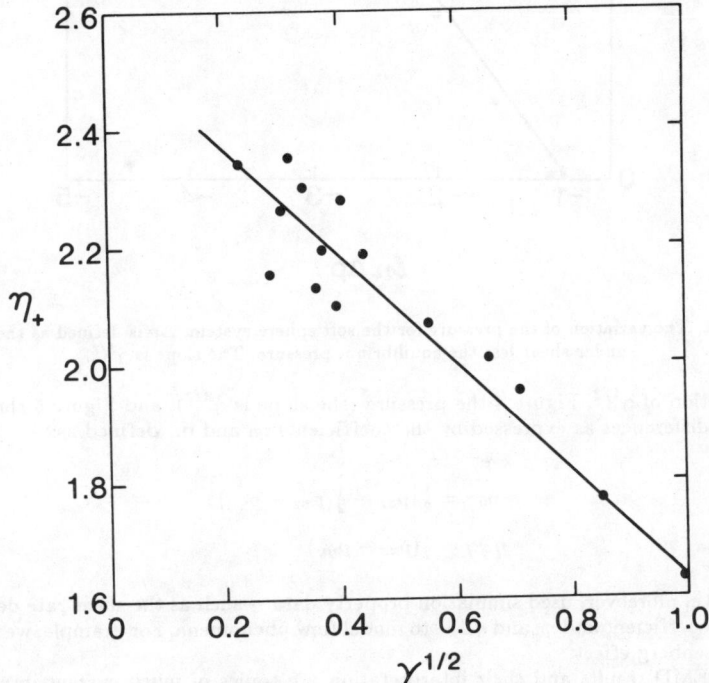

FIGURE 3. The viscosity of a soft-sphere liquid for a 108 particle r^{-12} system, extracted from reference 2. All variables are in reduced units with the reducing parameters set equal to one.

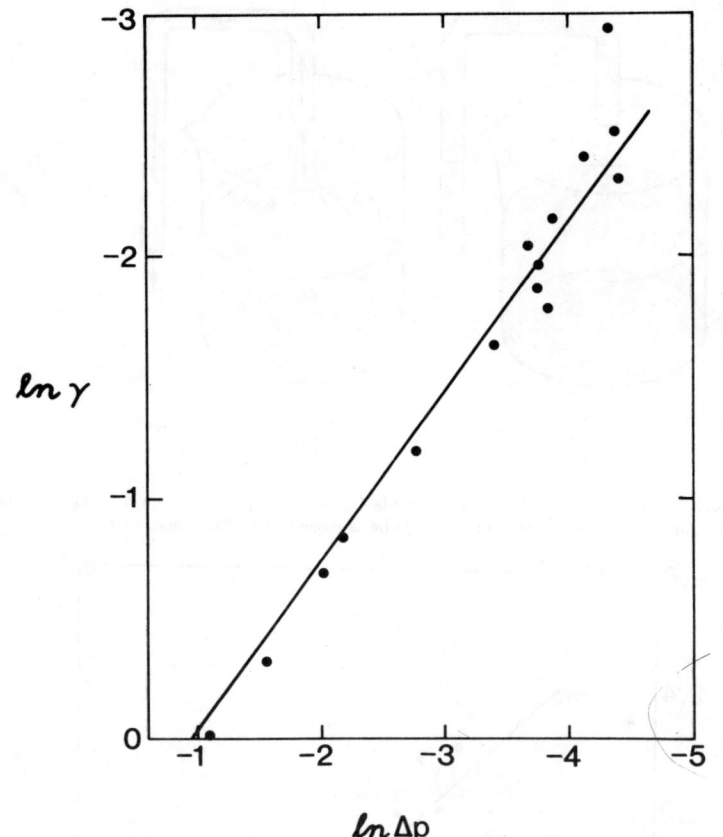

FIGURE 4. The variation of the pressure for the soft sphere system. Δp is defined as the pressure under shear less the equilibrium pressure. The slope is $\gamma^{3/2}$.

as a function of $\gamma^{1/2}$, Figure 4 the pressure (the slope is $\gamma^{3/2}$), and Figure 5 the normal pressure differences as expressed by the coefficients η_0 and η_- defined as:

$$\eta_0 \gamma = \tfrac{1}{2}\left[p_{zz} - \tfrac{1}{2}(p_{xx} + p_{yy})\right]$$
$$\eta_- \gamma = \tfrac{1}{2}(p_{zz} - p_{yy})$$
(3)

We have, moreover, used simulation property data —such as the shear rate dependent viscosity coefficient and η_0 and η_-— to model flow phenomena. For example, we modeled the Weissenberg effect[3]

The NEMD results and their interpretation are topics of much current interest and have engendered debate on the reliability of the computer algorithms, the underlying statistical mechanics of a system far from equilibrium under shear, and the relationship between the NEMD data and experiments. We prefer not to discuss these points here

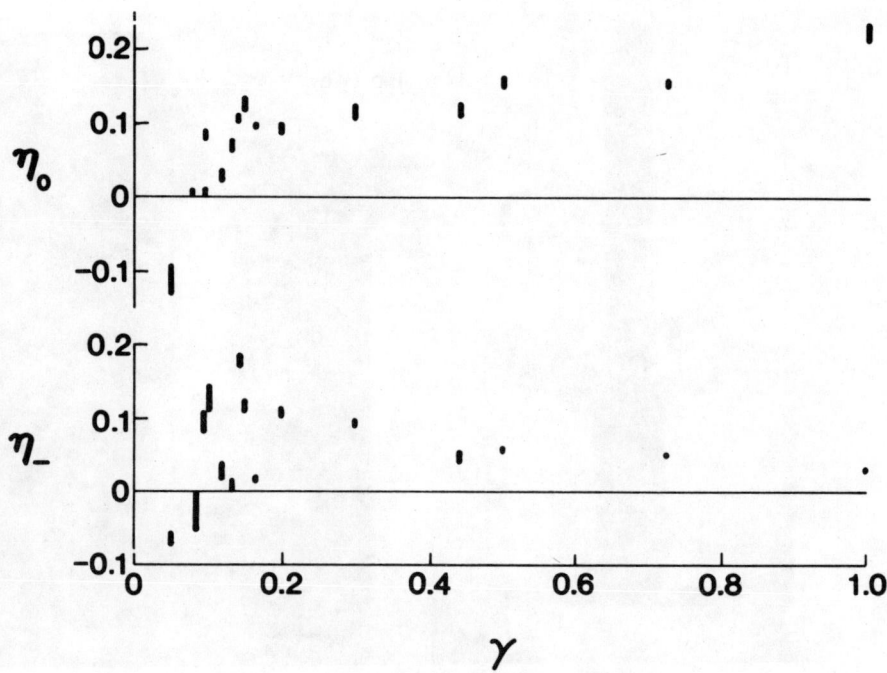

FIGURE 5. Plot of the normal pressure difference coefficients, equation (3), versus shear rate for the soft sphere system.

and refer to references 4 to 7 for details. But we can make two conclusions at this stage: 1) Clearly, then, one cannot simply distinguish Newtonian and non-Newtonian in terms of the molecular structure. In fact, the simulation results show that a Newtonian liquid is only a convenient approximation, similar in spirit to the ideal gas. It is necessary to introduce a parameter common to all liquids if one wants to estimate the experimental conditions under which a fluid will show-Newtonian characteristics. 2) Contrary to a widespread belief, these characteristics cannot be due solely to shear induced changes in intramolecular structure (such as the uncoiling of a polymer chain), distortions in the intermolecular microstructure must play a vital role.

3. A UNIVERSAL PARAMETER: THE MAXWELL RELAXATION TIME

A logical universal parameter to consider is the Maxwell relaxation time, τ, defined as the ratio of the shear viscosity to the shear modulus: $\tau = \eta_+/G$. One can also think of τ in terms of a diffusion coefficient, D, and a characteristic length, ℓ; specifically $\tau \sim \ell^2/D$. (The definitions are equivalent since G is an effective pressure and D relates to the viscosity according to the Einstein definition that $D = kT/m\xi$, where $\xi = 6\pi\ell\eta$.)

As a rule of thumb, "simple" liquids have relaxation times of the order of 10^{-12} to 10^{-14} s, while for "complex" liquids such as polymers $\tau = 10^{-4}$s or much greater. Note the concept of universality inherent in the definition of τ: τ can be large or small depending

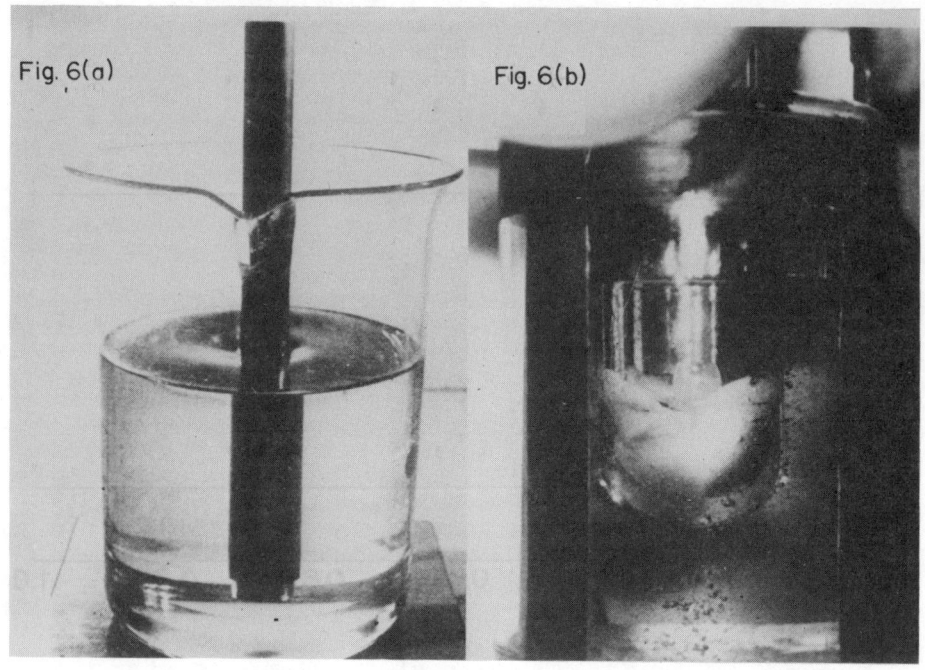

FIGURE 6. The stirring of glycerol, (a) at room temperature, (b) at 200 K. Note some evidence of a Weissenberg effect (rod climbing). See Figure 1.

on the relative magnitudes of η and G or ℓ and D and these magnitudes depend as much on the experimental conditions as on the nature of the molecule. Further, one of these conditions is the experimental observation time. Strictly one should redefine the concept of simple and complex with this in mind: a complex liquid is associated with relaxation time of the order of or larger than our observations.

The magnitude of τ is the common denominator between the simple and complex liquid. More precisely it appears that if the product $\tau\gamma$ in a sheared fluid is greater than about 10^{-2}, then one can expect (in the sense that one can observe) *any* fluid to display the long range collective non-Newtonian, nonlinear behavior. We note that, τ is of the order of 0.1 or 0.2 in the reduced units of the computer simulations for model systems close to their freezing density, and the shear rate is chosen to be between 0.01 and 1.0.

Behavior of glycerol

We have carried out a crude test of the relaxation time hypothesis. Let us consider the Weissenberg effect again. According to the hypothesis, a liquid with $\tau = 10^{-12}s$ will not show rod climbing on stirring under normal conditions, but if τ could be lengthened to 10^{-4}s or greater, the Weissenberg effect should be seen for a laboratory accessible shear rate. We carried out some very crude qualitative experiments with glycerol. Glycerol is remarkable in that its Maxwell relaxation time can change by many orders of magnitude, from 10^{-12} to 100s or greater, because it can be supercooled to 100 K below its nominal freezing temperature. Figure 6 demonstrates what happens if glycerol is stirred with a shear rate of $\sim 50s^{-1}$: 6 (a) is the expected result at room temperature $(\tau = 10^{-12}s)$, and 6 (b) the result at 200 K $(\tau = 10^{-3}s)$. There is crude evidence of rod climbing in 6 (b).

4. THE STRUCTURE FACTOR OF A TWO-DIMENSIONAL LIQUID

In the second part of the paper we discuss an evaluation of $S(\mathbf{k}, \gamma)$. As remarked in the introduction, the shear induced anisotropy in the structure factor must provide a key to understanding non-Newtonian behavior of all liquids. We have recently examined $S(\mathbf{k}, \gamma)$ for a two-dimensional soft disk system of 896 particles at constant density, ρ, and temperature, T, by NEMD[8] It is stressed that in this context there is no real loss in generality in working in two dimensions: A two-dimensional liquid is much simpler to work with than the three dimensional counterpart, and the results are much simpler to interpret. In fact we pointed out in reference 8 that several problems that clouded previous studies[9] of $S(\mathbf{k}, \gamma)$ have been clarified.

The simulation parameters are outlined in Figure 7. The liquid was studied at a density of 0.9238, which is 9/10 of the freezing density, and at a temperature of 1. (All variables are in reduced units and the energy length and mass parameters set equal to 1.) The potential of $\Phi = r^{-12}$ was truncated at $r = 1.5$. Stationary Couette flow was simulated via the usual NEMD technique with sliding periodic boundaries and an imposed shear rate, γ, defined as $\partial u_x/\partial y$ where u_x is the streaming velocity. Results were reported[8] for $0.1 \leq \gamma \leq 10$. Since τ for the system is about 0.15, the range of $\tau\gamma$ is $0.015 \leq \tau\gamma \leq 1.5$ —recall that one expects non-Newtonian characteristics if $\tau\gamma$ is in this range. In this paper we discuss some results for $\gamma = 1.0$, or for $\tau\gamma = 0.15$.

The NEMD algorithm is known as the SSLOD technique and has been described by Evans and Morriss[10] The algorithm has the powerful advantage that the kinetic temperature of the systems is kept constant by a Gaussian thermostat.

The equations of motion for the isokinetic SLLOD algorithm are

$$\dot{\mathbf{r}}_i = \frac{\mathbf{p}_i}{m} + \mathbf{n}_x \gamma y_i$$
$$\dot{\mathbf{p}}_i = \mathbf{F}_i - \mathbf{n}_x \gamma p_{yi} - \alpha \mathbf{p}_i \quad (4)$$

where $\mathbf{n}_x$ is a unit vector in the x direction, $\mathbf{r}_i$ is the position of particle i, $\mathbf{p}_i$ is the peculiar momentum and $\mathbf{F}_i = -\sum_j \partial \Phi_{ij}/\partial r_i$, and α is the Gaussian thermostatting

THE SYSTEM.

1. 896 particle 2-d liquid at $9/10$ of the freezing density.

2. $\phi = 1/r^{12}$

3. System undergoes couette flow

   $\gamma = \dfrac{du_x}{dy}$

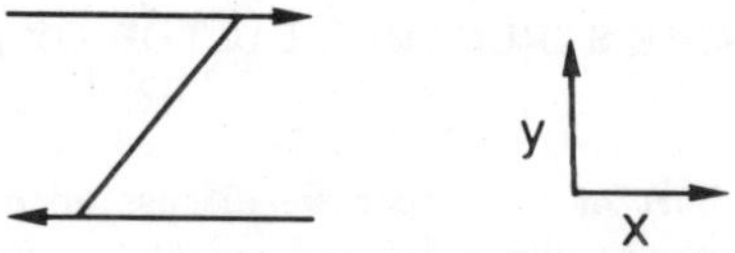

4. NEMD. Density constant, kinetic temperature constant via Gaussian thermostat.

   Range of γ such that
   $0.1 \leq \tau\gamma \leq 10$

   [with $\tau \approx 0.15$]

FIGURE 7. Parameters for the NEMD simulation.

multiplier given by

$$\alpha = \frac{\sum^N{}_{i=1}(\mathbf{p}_i \cdot \mathbf{F}_i - \gamma p_{xi} p_{yi})}{\sum_{i=1}^{N} p_i^2} \qquad (5)$$

The Gaussian thermostat fixes the value of the kinetic temperature by setting its time derivative to zero. The SLLOD algorithm, together with the displaced periodic boundary conditions, is the exact algorithm for implementing a linear velocity profile with periodic boundary conditions. Given a stable linear velocity profile the definition of the kinetic temperature is as usual, and thus the SLLOD algorithm combined with the Gaussian gives a well defined statistical mechanical idealization of Couette flow, in the same sense as the microcanonical, canonical and grand canonical ensembles are valid representations of equilibrium systems. The technical details of the algorithm are standard.[10] Equations (4)

FIGURE 8. The structure factor of a two dimensional soft disk liquid under a shear rate of 1.0, obtained by scattered light from a medium constructed from the coordinates of 896 particles in Couette flow.

are solved using a fifth order Gear predictor-corrector scheme with a reduced time step of 0.004.

The structure factor

The structure factor was evaluated directly. We established the steady state with $\gamma = 1.0$. The simulation was then restarted from the steady state, then stopped after a random number of time steps. The co-ordinates of the 896 particles were then represented as dots onto photographic film. The film scale chosen was such that the simulation box length was represented by 25 mm. The dot size was 200 μm[11] (chosen for convenience from a limited number of standard sizes available). The dots were positioned on the film with an accuracy of 0.05%. The film with the dots was then treated as a scattering medium and the diffraction pattern [*i.e.*, $S(\mathbf{k}, \gamma)$] measured with a low power He/Ne laser.

Although optical transforms were readily obtained from a single set of 896 co-ordinates, the patterns obtained were quite noisy because of the inherently poor statistics. Consequently to improve the quality of the transforms, diffraction masks were prepared which consisted of a composite plot of sixteen sets of data taken randomly at various times during the same NEMD simulation. These were arranged side by side on a 4×4 grid covering an area of 10 cm × 10 cm on the film. Since there can be virtually no phase coherence between light scattered from the separate plots the resulting intensity can be assumed to be the average of the individual transforms.

The structure factor from this method is shown as Figure 8. One sees the characteristic ellipse of the diffuse rings with the major axis of the ellipse at approximately $3\pi/4$. However, as we will discuss in the next section the axis is not quite at $3\pi/4$ indicating a small normal pressure difference in the fluid.

5. EXPANSION OF THE PAIR CORRELATION FUNCTION

The microstructure of the fluid can be approached through a study of the pair correlation function. This is, of course, standard if the fluid is at equilibrium when the pair correlation function is the equilibrium radial distribution function. We have discussed the pair correlation function for a system under shear, $g(\mathbf{r}, \gamma)$, in reference 9 and references therein.

Because the pair correlation function $g(\mathbf{r}, \gamma)$ in polar coordinates is periodic with period 2π in two dimensions,[8] we can expand it as a single variable Fourier series. The structure is then represented by some finite resummation of the series

$$g(\mathbf{r}, \gamma) = g(r, \theta, \gamma)$$
$$= g_s(r, \gamma) + \sum_{n=1}^{\infty} \left[g_0^{(n)}(r, \gamma) \sin n\theta + g_1^{(n)}(r, \gamma) \cos n\theta \right] \quad (6)$$

where

$$g_s(r, \gamma) = \frac{1}{2\pi} \int_0^{2\pi} d\theta \, g(r, \theta, \gamma) \quad (7)$$

$$g_0^{(n)}(r, \gamma) = \frac{1}{\pi} \int_0^{2\pi} d\theta \, g(r, \theta, \gamma) \sin n\theta \quad (8)$$

$$g_1^{(n)}(r, \gamma) = \frac{1}{\pi} \int_0^{2\pi} d\theta \, g(r, \theta, \gamma) \cos n\theta \quad (9)$$

For convenience we exclude the γ-dependence in the subsequent notation. Since Couette flow is symmetric with respect to rotation through π, all coefficients with odd values of n are zero.

Appropriate integrals of $g_s(r)$ give the potential contribution to the shear rate dependent thermodynamic properties of the system, for example, the energy E_Φ is given by,

$$E_\Phi = \pi \rho \int_0^\infty dr \, r^{-11} g_s(r) \quad (10)$$

IN GENERAL:

$$g(\mathbf{r}) = g_s(r) + \sum_n^\infty [g_0^{(n)}\sin n\theta + g_1^{(n)}\cos n\theta]$$

EQUILIBRIUM:

$$g(\mathbf{r}) \longrightarrow g_s(r) \longrightarrow g_{eq}(r)$$

NEWTONIAN:

$$g(\mathbf{r}) = g_{eq}(r) + g_0^{(2)} \sin 2\theta$$

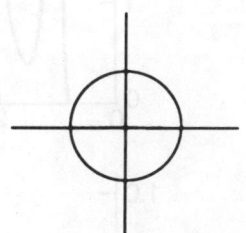

NON NEWTONIAN:

$$g(\mathbf{r}) = g_s(r) + g_0^{(2)} \sin 2\theta + g_1^{(2)} \cos 2\theta + \ldots$$

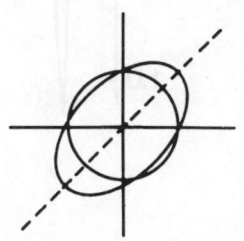

FIGURE 9. Relationships between the pair correlation function, $g(\mathbf{r},\gamma)$ and the structure factor. See equation (6), and the discussion in the text.

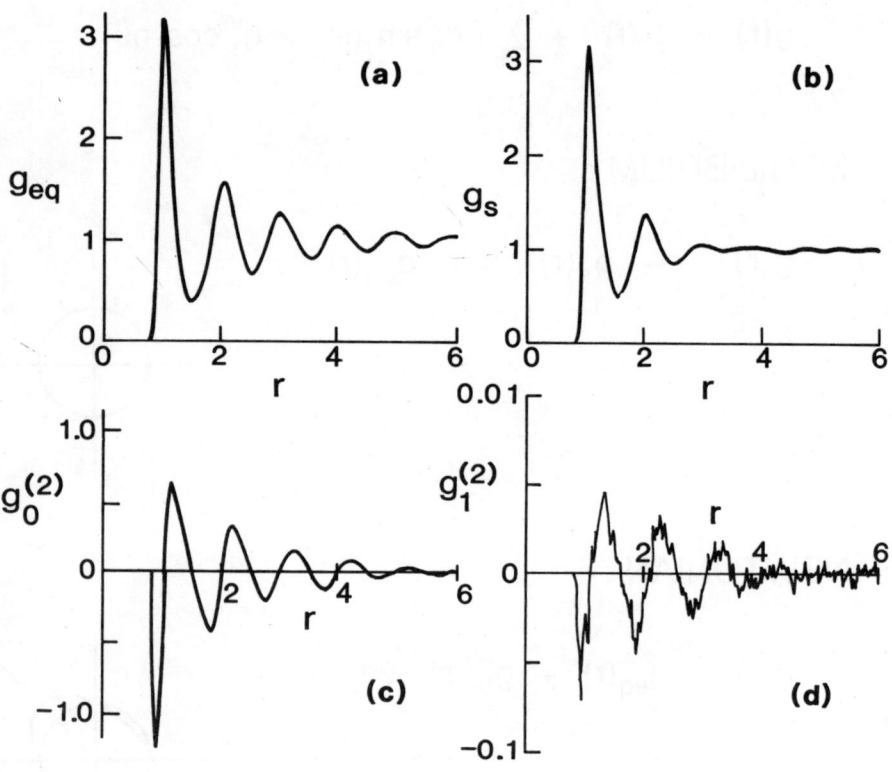

FIGURE 10. Plot of the equilibrium radial distribution function for the 896 soft disk system at $\rho = 0.9238$; (a) Plot of the scalar contribution of equation (6) for the system at $\gamma = 1.0$; (b), (c) and (d) are plots of the $g^{(2)}$ coefficients.

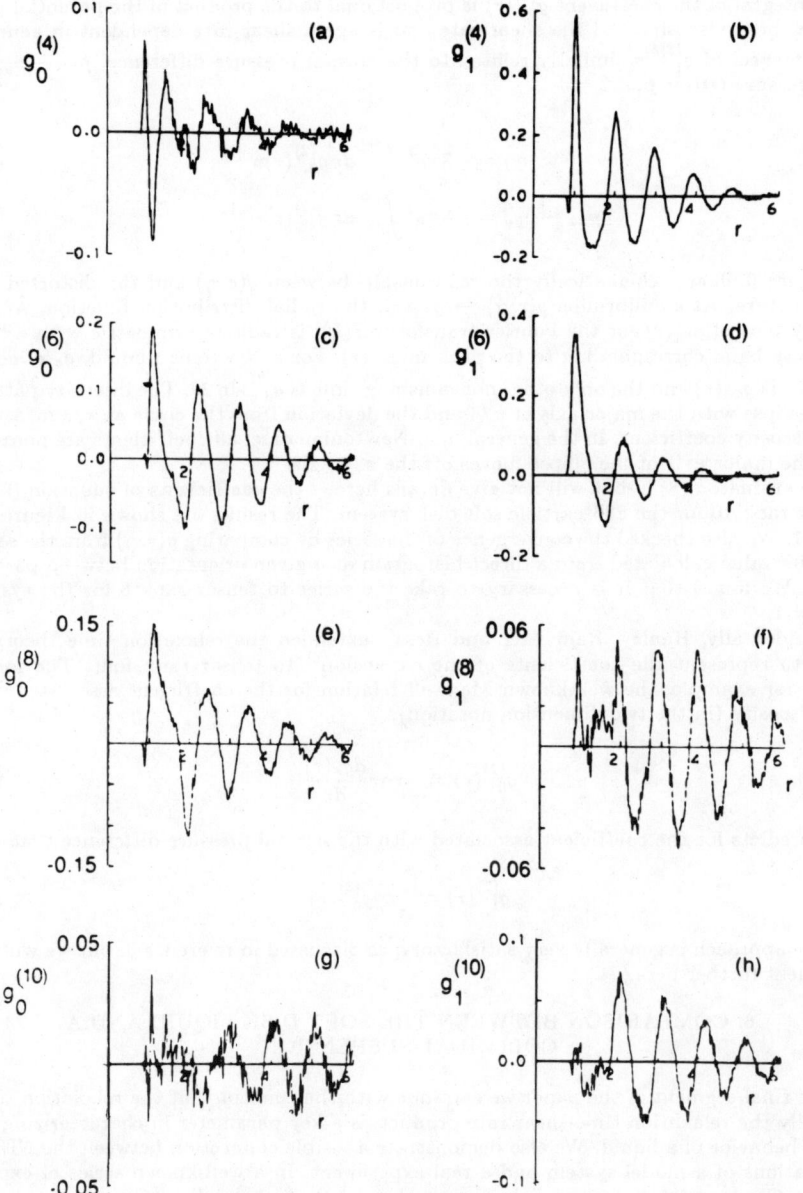

FIGURE 11. Plot of the $g^{(n)}$ coefficients with $4 \leq n \leq 10$ from equation (6) from the soft disk system with $\gamma = 1.0$.

The integral of the coefficient $g_0^{(2)}(r)$ is proportional to the product of the potential part of the shear viscosity and the shear rate and is again shear rate dependent in general. The integral of $g_1^{(2)}$ is similarly related to the normal pressure difference, $p_{xx} - p_{yy}$ of the pressure tensor $\underset{\approx}{p}$,

$$\eta\gamma = -3\pi\rho^2 \int_0^\infty dr\, g_0^{(2)}(r) r^{-11} \tag{11}$$

$$p_{xx} - p_{yy} = -3\pi\rho^2 \int_0^\infty dr\, g_1^{(2)}(r) r^{-11} \tag{12}$$

Figure 9 shows schematically the relationship between $g(\mathbf{r},\gamma)$ and the distorted microstructure. At equilibriums $g(\mathbf{r},\gamma) \to g_{eq}(r)$, the radial distribution function. An intensity plot of $g_{eq}(r)$ [or the Fourier transform $S(\mathbf{k})$] is radially symmetric with a high intensity band corresponding to the peak in $g_{eq}(r)$. For a Newtonian liquid, g_s of equation (6) is $g_{eq}(r)$ and the only other nonvanishing time is $g_0^{(2)} \sin 2\theta$. The intensity pattern is an ellipse with the major axis at $\pi/4$ and the deviation from the circle at $\pi/4$ measures the viscosity coefficient. In the general, non-Newtonian case, all coefficients are nonzero, and the major axis of the ellipse moves off the $\pi/4$ axis.

We evaluated (8) —but will not give details here— the coefficients of equation (6) to tensor rank 10 for the 896 particle soft disk system. The results are shown in Figures 10 and 11. We also checked the convergence of the series by comparing $g(\mathbf{r},\gamma)$ from the series with its value calculated from a direct histogram at a given orientation between particle pairs. We found that it is necessary to take the series to tensor rank 8 for the system at $\gamma = 1$.

[Incidentally, Hanley, Rainwater and Hess[8] extended the relaxation time theory of Hess to represent the coefficients of the expansion[6] to tensor rank four. The model gives, for example, the well-known Maxwell relation for the coefficient associated with the viscosity (in the two-dimension notation);

$$g_0^{(2)}(r) = -\tau\gamma r \frac{dg_s(r)}{dr} \tag{13}$$

and predicts for the coefficient associated with the normal pressure difference that

$$g_1^{(2)}(r) = -\tau\gamma g_0^{(2)}(r) \tag{14}$$

The approach is generally very satisfactory, as discussed in reference 8, but we will not comment further here.]

6. COMPARISON BETWEEN THE SOFT DISK LIQUID AND A COLLOIDAL SUSPENSION

In the final segment of the paper we continue with the concept that the relaxation time, actually the relaxation time-shear rate product, is a key parameter in characterizing the shear behavior of a liquid. We also demonstrate a visible connection between the NEMD simulations of a model system and a real experiment. In a well-known series of experiments, Clark and Ackerson and their co-workers[12] studied the distorted microstructure of colloidal suspensions under shear via light scattering. The relaxation time of the colloids is $= 10^{-2} s$ and the $\tau\gamma$ product of their experiments is $= 0.1$, comparable to the NEMD conditions.

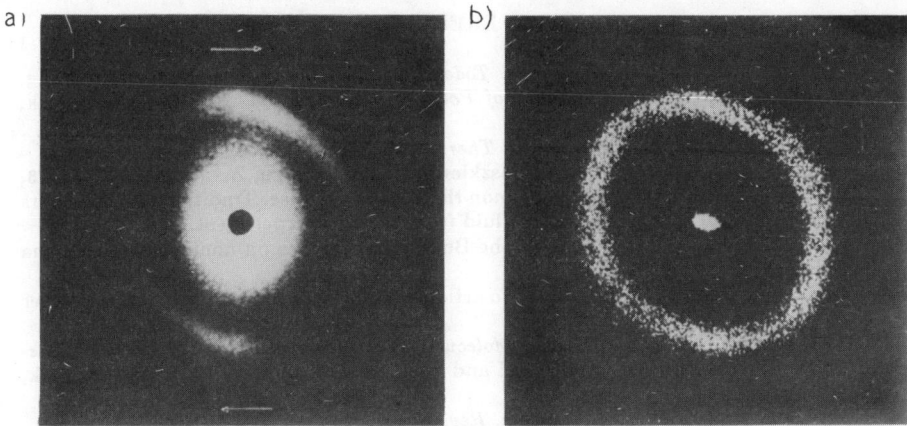

FIGURE 12. (a) Scattered light intensity from a real colloidal suspension undergoing shear. Reproduced from Hanley, et al. reference 13. (b) The Optical Transform from the NEMD simulation as in Figure 8.

Figure 12 (a) here is a reproduction of Figure 1 (b) from reference 13 and shows the colloidal structure factor; note the high intensity spots just off the vertical. [The central black spot and the scattered light around the centre spot are artifacts of the experiment.] Figure 12 (b) was constructed from the identical information used in Figure 8 for $\gamma = 1.0$, only the background intensity has been adjusted to emphasize the inner intensity ring. The optical transform for the NEMD example shows great similarity to the structure factor for the real colloid since it shows the characteristic elliptic shape of the diffuse peak and moreover, this contains clearly visible bright spots. [The optical patterns clearly show a second outer diffuse ring, however, which is not apparent in the patterns for the real colloid. Use of a somewhat larger aperture in our optical experiment would diminish the relative intensity of this peak, but it is also possible that the lack of perfect monodispersion of the particle size in the real colloid system might cause this and higher order peaks to be smeared out.]

7. SUMMARY AND CONCLUSIONS

This paper has reviewed very briefly the evidence that simple fluids are in principle non-Newtonian. We have argued that it is practical to consider the relaxation time, τ, of a fluid and to classify fluids in terms of τ rather than molecular structure. We have also argued that it is productive to study the distorted microstructure of a sheared liquid via the scattered intensity or the structure factor, $S(\mathbf{k}, \gamma)$. A novel way to obtain $S(\mathbf{k}, \gamma)$ directly from an optical transform of results from nonequilibrium molecular dynamics has been described and the results compared to an experimental colloidal suspension.

Much of this work is based on extensive discussions with J.C. Rainwater and S. Hess and follows closely a paper written with G.P. Morriss, T.R. Welberry and D.J. Evans at the Australian National University. The work was supported in part by the Department of Energy, Office of Basic Energy Sciences.

REFERENCES

1. Bird, R.B. and Curtiss, C.F., *Phys. Today* **36** (Jan. 1984); Bird, R.B., Armstrong, R.C. and Hassager, O., *Dynamics of Polymeric Liquids Vol. 1* (Wiley, New York, 1977)
2. Hess, S. and Hanley, H.J.M., *Int. J. Thermophys.* **4**, 97 (1983).
3. Rainwater, J.C., Hanley, H.J.M., Paszkiewicz, T. and Petru, Z., *J. Chem. Phys.* **83**, 339 (1985). This is a reversal of common rheological practise. Traditionally, one infers the properties of a non-Newtonian fluid from the flow experiment.
4. *Physica* **A118** (1983). Reported on the Boulder conference on nonlinear phenomena in fluids.
5. *Phys. Today* (Jan. 1984). Is devoted to articles on non-Newtonian behavior, computer simulation and kinetic theory.
6. Hanley, Holian, Dufty and Evans, *Molecular-dynamics simulations of statistical-Mechanical Systems*. Ed. by G. Ciccotti and W.G. Hoover (North-Holland, New York, 1986).
7. Evans, D.J. and Hoover, W.G., *Ann. Rev. Fluid Mech.* **18**, 243 (1986).
8. Hanely, H.J.M., Morriss, G.P., Welberry, T.R. and Evans, D.J., *Physica* (in press).
9. Hanely, H.J.M., Rainwater, J.C. and Hess, S., *Phys. Rev.* **A36**, 1795 (1987).
10. Evans, D.J. and Morriss, G.P., *Comput. Phys. Rep.* **1**, 297 (1984).
11. Provided that no overlapping occurs, the actual size of the aperture effects only the overall variation of scattered intensity with [k], since convolution of the aperture shape with the point distribution of particles in real space, results in a simple multiplicative factor in k-space.
12. Clark, N.A. and Ackerson, B.J., *Phys. Rev. Lett.* **44**, 1005 (1980); Clark, N.A., Ackerson, B.J. and Hurd, A.J., *Phys. Rev. Lett.* **50**, 1459 (1983).
13. Hanley, H.J.M., Rainwater, J.C., Clark, N.A. and Ackerson, B.J., *J. Chem. Phys.* **79**, 4448 (1983).

125

TRANSIENT PATTERN FORMATION IN NONEQUILIBRIUM FLUIDS

Rashmi C. Desai and Kenneth R. Elder

Department of Physics, University of Toronto,
Toronto, Ont, M5S 1A7, CANADA

ABSTRACT. The relaxation to single phase equilibrium of a one component two dimensional Lennard-Jones fluid following *large* quenches is investigated using the constant-density molecular dynamics technique. Simulations corresponding to constant energy and constant temperature are performed and compared with fluctuating linear hydrodynamics. After three or four correlation times, during which equilibrium in momentum space is established, there is a good agreement between the theory and simulation. The large-quench, single-phase simulation results are discussed in relation to similar simulations which result in phase separation.

1. INTRODUCTION

Development of inhomogeneities and consequent pattern formation is a common occurrence in a variety of nonequilibrium phenomena. It is seen vividly during the phase separation of binary mixtures of fluids, metals, polymers etc. and its understanding is an important aspect of the current research in the dynamics of first order phase transition.[1] Theory, experiment and computer simulations[2] are being used in a complementary way to enhance our understanding of this rich and complex area.

In the computer simulations, phase separation dynamics is typically studied in the following way. The system is prepared in a homogeneous equilibrium state within the single phase region; it is then rapidly quenched (in one time step) to a nonequilibrium state within the two phase coexistence region: one changes the size of the system, keeping the relative interparticle distances fixed, in order to change its density and one changes all the particle velocities by a constant factor in order to change the system temperature.[3] Once the initial post-quench state is thus prepared, its subsequent time evolution is monitored to obtain the simulation data related to the pattern formation and other aspects of the dynamics of the system as it phase separates. Similar quenches can also be performed where the initial post-quench state is such that the system does not phase separate but escapes back to the single phase region and eventually equilibrates to a homogeneous single phase state, which may be far from the initial pre-quench state. During its time evolution following such a quench, the system displays interesting transient patterns, even though both the initial and infinite time states are homogeneous. In this paper, we present some new results related to such quenches.

We have also done an extensive study of time evolution of quenched systems which do phase separate.[4] Most of these later simulations are such that the initial post-quench state of the system is thermodynamically unstable and the subsequent phenomena is usually referred to as spinodal decomposition[1, 5, 6]. During the process, typically spaghetti like patterns develop and often span the system (in our case, a two-dimensional box with periodic boundary conditions); these spatial inhomogeneities in the order parameter Ψ evolve into macroscopic domains and eventually into a two phase coexistence. The phenomena of spinodal decomposition is complex, but rich in physics. It involves (i) instability, (ii) irreversibility, (iii) dynamic creation of interfaces, and (iv) growth of ordered domains. During the irreversible approach to the eventual phase separated state, conservation laws also play a crucial role in the dynamics. In critical dynamics, one has model classifications depending on the conservation laws: model A-Ψ not conserved, model B-Ψ conserved, and so on. The simulations in reference 4 are for the Langevin equation appropriate to model B: it is a stochastic nonlinear partial differential equation in time and the 2-d space. In this paper, we present the results for a one-component, two-dimensional Lennard-Jones fluid and we use the molecular dynamics (MD) simulation technique, for details regarding MD simulation technique, see reference 7. For a one component fluid, the order parameter Ψ is the fluid density ρ. It is conserved, and is coupled to the momentum and energy densities which are also conserved. It is the liquid-vapour phase separation that is of relevance. For the quench results presented here in which the system goes back to the single phase region, the irreversibility and conservation laws are the features in common with the phase separation dynamics.

A variety of questions have been studied in literature which characterize the dynamics of first order phase transition. These include: (i) time dependent structure factor $S(\mathbf{k},t)$ and its scaling properties in relation to the growing domain size $R(t)$ and to other less important lengths (e.g. domain wall width) during the phase separation process, (ii) time dependence of $R(t)$ and the associated power law growth exponent x, (iii) time dependence of the excess internal energy density and its power law decay exponent y: is $y = (1 - 2x)$? what is the connection to the thermodynamic latent heat, etc. (iv) cluster size distribution of the ordered domains and its time evolution, (v) details of the system morphology whether growing clusters are compact of fractal and whether there is a percolating network, (vi) range of validity of linear and mean field like theories.

For the large perturbation quenches which bring the system to the single phase, dynamics can be explored through the structure factor. We compute the $S(\mathbf{k},t)$ and compare the simulation results with the linear fluctuating hydrodynamics. The purpose of such a comparison is to assess the validity of the macroscopic linear theory for a simpler case (compared to phase separation) of approach to single phase equilibrium following a far from equilibrium quench.

2. STRUCTURE FACTOR

The time dependent structure factor $S(\mathbf{k},t)$ is the equal time correlation function of the fluctuation in the order parameter:

$$S(\mathbf{k},t) = \langle \rho(\mathbf{k},t) \rho^*(\mathbf{k},t) \rangle_{\text{ne}} / N, \tag{1a}$$

where $\rho(\mathbf{k},t)$ is the Fourier component of the order parameter fluctuation,

$$\rho(\mathbf{k},t) = \int d^2 r \, e^{-i\mathbf{k}\cdot\mathbf{r}} [\rho(\mathbf{r},t) - \rho_0], \tag{1b}$$

with ρ_0 being the mean density [which also is the mean homogeneous density in the pre-quench equilibrium state] and N being the number of particles in the system [$N = 5184$ in our simulations]. In equation (1), the angular brackets stand for the average over the nonequilibrium state (distribution function) of the system. For systems in equilibrium or in steady state, $S(\mathbf{k},t)$ becomes time independent and is nothing but the usual structure factor $S(\mathbf{k})$ which is measured by X-ray and neutron diffraction experiments.[9] In our quenches, the system would approach a new single-phase equilibrium state (ρ_0, T_0) at $t = \infty$ and $S(\mathbf{k},t)$ would approach $S(\mathbf{k})$ appropriate to this state.[10]

We compare the simulation results to the predictions of fluctuating linear hydrodynamic (FLH) theory[7,8] in order to see the limitations of a linear theory. The theoretical derivation of $S(\mathbf{k},t)$ based on FLH is described in detail in reference 8 and the result is given in its equation (2.18). The basic idea is to linearize the equations of fluctuating hydrodynamics (*i.e.* the mass balance, Navier-Stokes [including fluctuation in the stress tensor] and energy balance [including fluctuations in the heat flux] equations) and then solve for $\rho(\mathbf{k},t)$ by a Fourier-Laplace transformation. The pressure gradient term appearing in the Navier-Stokes equation is related to the temperature and density fluctuation in the usual way and a term related to the van der Waals square-gradient free energy is also added to take into account the nonlocal correlations (equations 2.8 and 2.22 in reference 8). If one assumes that the initial values of temperature and velocity fluctuations are small and can be ignored, then $\rho(\mathbf{k},t)$ depends only on its own initial value. Once $\rho(\mathbf{k},t)$ is obtained, it can be used to obtain $S(\mathbf{k},t)$. The final result depends on various transport coefficients and thermodynamic derivatives which are computed using the Enskog theory of dense fluids,[11] Monte Carlo data[12] on Lennard-Jones 2-d fluids, and the van der Waals equation of state. The use of more realistic equation of state and a generalized nonlocal free energy does not yield any qualitative changes[7,8] for small wavenumbers.

3. NUMERICAL EXPERIMENTS

In the MD simulations, the one evolution of 5184 Lennard-Jones particles in two dimensions was determined by numerically solving Newton's equation of motion using a fifth order Nordsiek-Gear corrector-predictor algorithm. The simulations being at constant density involved the particles moving in a box of fixed size with periodic boundary conditions. The phase diagram of the system is known[13] and is also shown in Figure 1 of reference 7. The critical point (ρ_c, T_c) is at $(0.325, 0.56)$. To facilitate performing the desired quenches, a pre-quench equilibrium state was obtained at $(\rho, T) = (0.8, 1.0)$ in the single phase region. The time step[10] in the simulation was 0.005 and all quenches were performed in one time step. We designed the pre-quench state to be such that it had a rather low liquid-like compressibility, and performed quenches to various other states (ρ_i, T_i) such that the eventual equilibrium state (ρ_0, T_0) would have a higher compressibility. Various initial post-quench states (ρ_i, T_i) for which the simulations were performed are $(0.765, 0.5)$, $(0.765, 0.836)$, $(0.325, 0.01)$, $(0.325, 0.6)$, $(0.325, 1.0)$, $(0.325, 1.2)$, $(0.6, 0.7)$, $(0.25, 1.0)$ and $(0.325, 0.7)$. Both constant temperature —CT— (using velocity rescaling) and constant energy simulations —CE—[8] were carried out. We single out among these simulations, the CT simulations following the quench to $(0.325, 0.7)$ and the CE simulation following the quench to $(0.325, 0.01)$. In the later simulation, the temperature rises to a value of about 0.71 at $t = 18$, with the equilibrium value estimated at 0.73. Thus it is interesting to compare the time evolution of these two simulations since their final equilibrium states are nearly the same.[14]

4. RESULTS

Figure 1 depicts snapshot configurations for these two simulations at various times. Visual examination of these pictures shows the growth of long wavelength density fluctuations. An earlier work by Fehder[15] on a system of 364 Lennard-Jones particles also revealed the presence of "holes" just above T_c. We find that this phenomena occurs in any situation

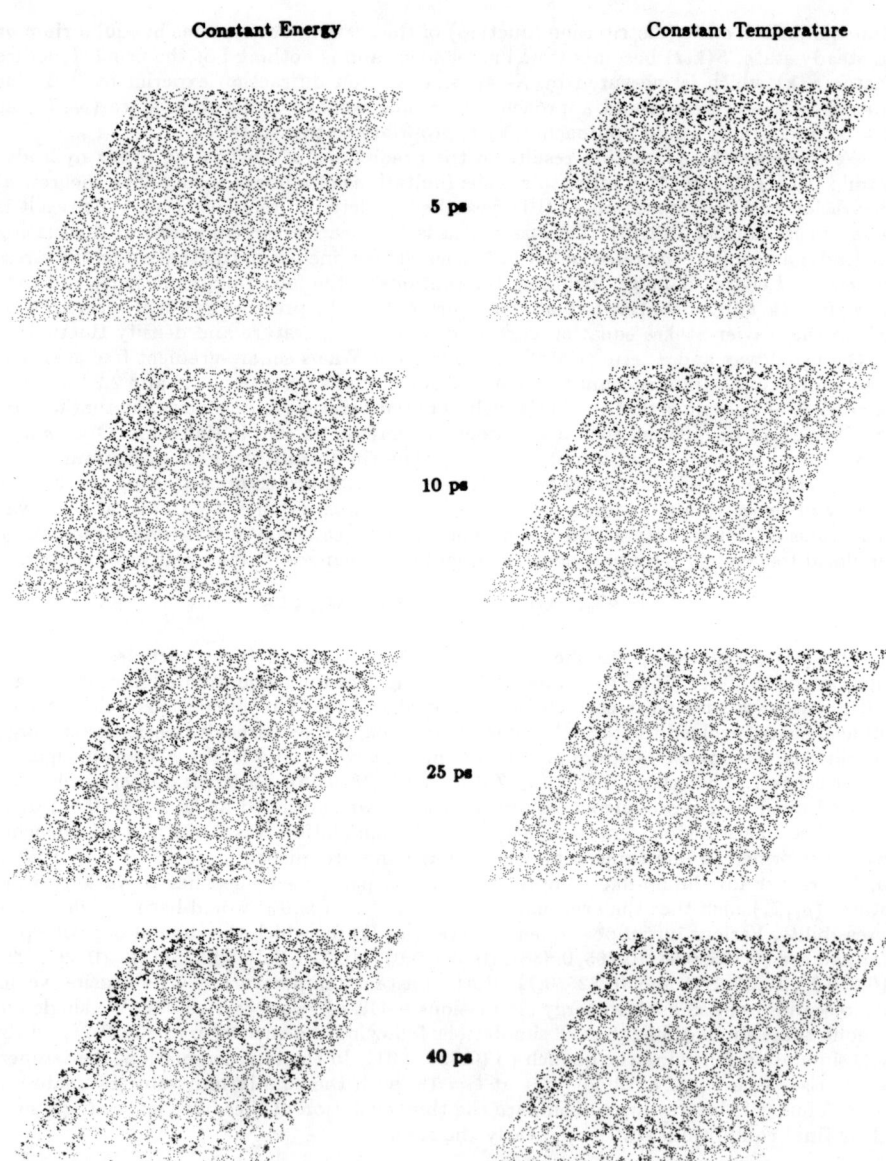

FIGURE 1. Snapshots of the time evolution of a 2-d Lennard-Jones fluid following a large-quench: Left half —$(\rho, T) = (0.8, 1.0) \rightarrow (0.325, 0.01) \rightarrow$ constant energy evolution. Right half —$(\rho, T) = (0.8, 1.0) \rightarrow (0.325, 0.7) \rightarrow$ constant temperature evolution using the velocity rescaling method.[7] See the remarks in reference 10 re. timescale.

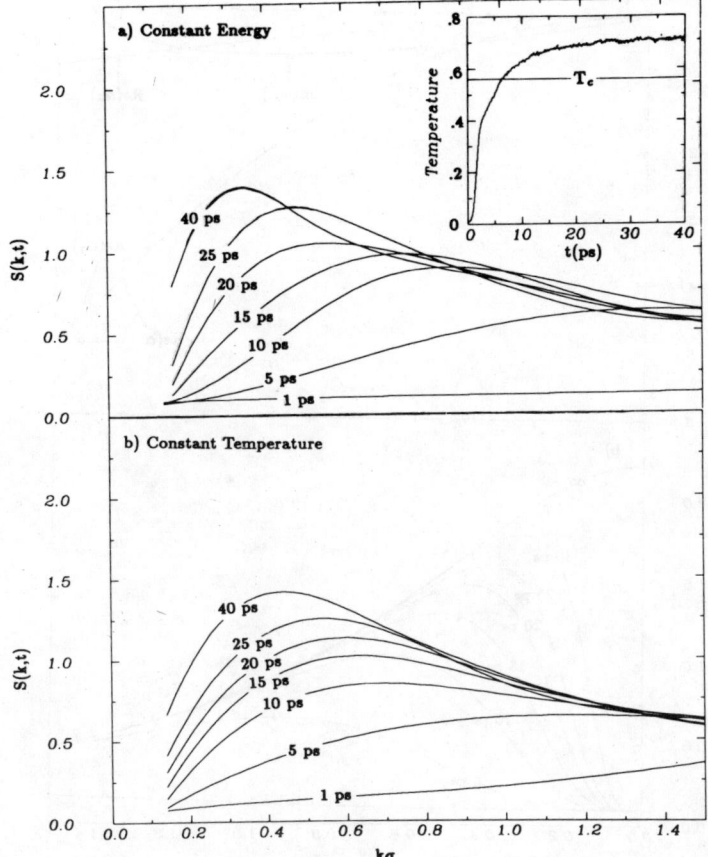

FIGURE 2. Time dependent structure factor as a function of wavevector at various times: (a) Constant energy simulation —inset shows the variation of global temperature with time; (b) Constant temperature simulation. See remarks in reference 10 re. timescale.

where the system is evolving towards a more compressible state, and is not restricted to the critical region.

In Figure 2, we show the structure factor $S(\mathbf{k}, t)$ as a function of wavenumber k, for various times. Figure 2 (a) shows it for the CE simulation, with the time variation of the "temperature"[3, 10)] shown in the inset; Figure 2 (b) shows it for the CT simulation. For the CE simulation, the initial post-quench state is deep within the two phase coexistence region; but by $t = 2.8$, the system has moved back to the single phase region and has reached a near-equilibrium value of $T = 0.7$ at $t = 10$. The main qualitative observation is the development of a peak in $S(\mathbf{k}, t)$ whose position moves to a smaller wavenumber with increasing time. This corresponds to the patterns seen in Figure 1. It should be clear that these patterns are transient for the nonequilibrium time evolution of the system. At infinite time, the peak position will approach $k = 0$, and the value of the structure factor at the peak will correspond to the isothermal compressibility χ: $\lim_{k \to 0} \lim_{t \to \infty} S(\mathbf{k}, t) =$

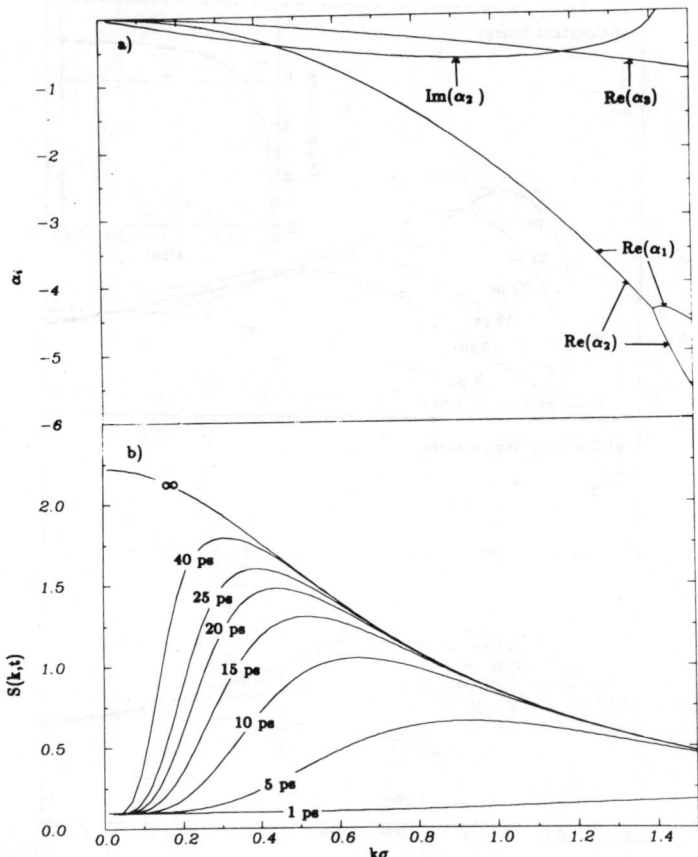

FIGURE 3. (a) Dispersion of hydrodynamic modes (sound modes α_1, α_2, and the thermal diffusivity mode α_3); (b) Structure factor from the FLH theory. (Compare to Figure 2 and see remarks in reference 10 re. timescale.)

$\rho_0 T_0 \chi$. In the above relation, the order of the limits is important and in the simulation with a box length of $L \ (= (N/\rho_0)^{1/2})$, the minimum accessible wavenumber is $2\pi/L$. With these remarks in mind, it is useful to compare the simulation results of Figure 2, with the prediction of FLH theory.

In Figure 3, we show the dispersion of hydrodynamic modes [Figure 3 (a)], and $S(\mathbf{k}, t)$ as functions of k for the FLH theory. The qualitative similarity between Figures 2 (a), 2 (b) and 3 (b) shows the predictive power of FLH. An interesting peculiarity of the FLH is that in the linear approximation, the global temperature T_g is time independent (*i.e.* the linearized energy balance equation implies that $\partial T_g/\partial t = 0$, where $T_g = \rho_0^{-1} V^{-1} \int T(\mathbf{r}, t) \rho(\mathbf{r}, t) \, d\mathbf{r}$). This in turn implies that the FLH predictions for the structure factor $S(\mathbf{k}, t)$ in the CE and CT simulations are identical if they have the same final equilibrium state (ρ_0, T_0). In this sense FLH is clearly deficient. The difference between the two simulations is clearly apparent at early times. During the first few correlation times, the system presumably equilibrates in the momentum space; and it is here that any

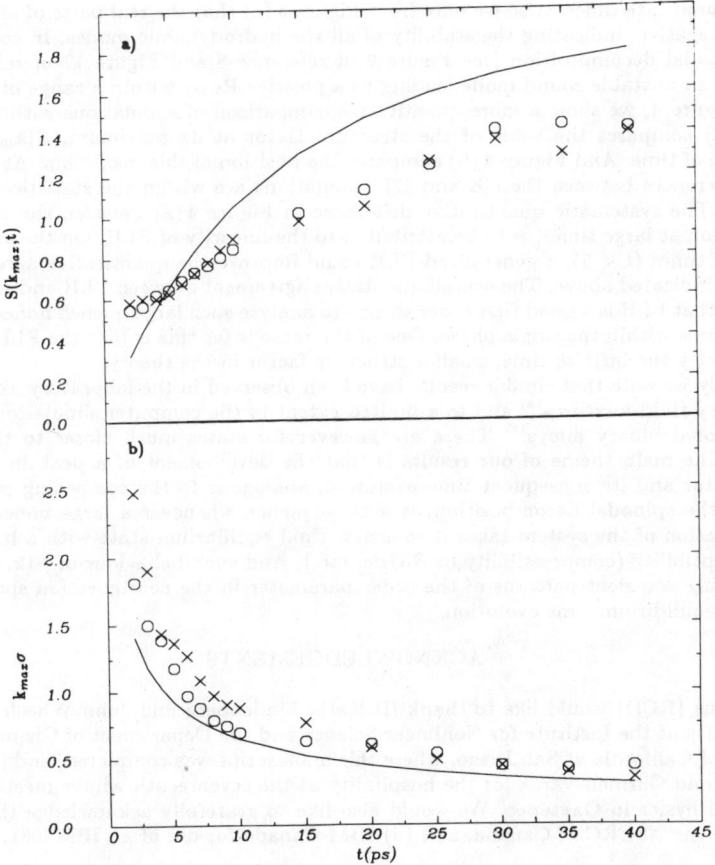

FIGURE 4. Comparison of numerical simulation and the FLH theory: (a) Value of the peak structure factor $S(\mathbf{k}_{max}, t)$; (b) Value of the peak position k_{max}, as functions of time. Solid line — Theory; open circles —Constant temperature simulation, crosses —Constant energy simulation. See remarks in reference 10 re. timescale.

hydrodynamic theory, linear or nonlinear, could be found wanting. In spite of this, the qualitative agreement between FLH and simulations is interesting. One of the reasons for the differences between theory and simulation is that the simulation curves are affected by the small-k wing of the first diffraction peak of $S(\mathbf{k}, t)$, which is associated with the short range liquid-like order. This short range order is of course not included in FLH and to obtain a quantitative agreement, one has to use a generalized hydrodynamic approach, particularly at short times and large wavenumbers, $k > \sigma^{-1}$. In such a generalized FLH, it is important to also include the feature [inset, Figure 2 (a)] that temperature is time dependent in the CE simulation. Another remark that should be made is the similarity and contrast of $S(\mathbf{k}, t)$ in these simulations with the analogous simulation results for spinodal decomposition[7] The coarsening phenomena in the later case implies unlimited growth of $S(\mathbf{k}, t)$ as t increases. In contrast, for the single phase evolution, $S(\mathbf{k}, t)$ reaches a plateau at late times. Also we note from Figure 3 (a) that the real parts of all the roots α_i are negative, indicating the stability of all the hydrodynamic modes. In contrast, for the spinodal decomposition (see Figure 2 of reference 8 and Figure 17 of reference 7), there is an unstable sound mode leading to a positive Re α_1 within a range of k values.

In Figure 4, we show a more quantitative comparison of simulations with FLH. Figure 4 (a) compares the value of the structure factor at its maximum, $S(\mathbf{k}_{max}, t)$ as a function of time. And Figure 4 (b) compares the position of this maximum. At late times the differences between the CE and CT simulations are within the statistical errors of $S(\mathbf{k}, t)$. The systematic quantitative difference in Figure 4 (a) between the theory and simulation at large times, is to be attributed to the linearity of FLH. On the other hand, at early times ($t < 5$), a generalized FLH could improve the quantitative agreement for reasons indicated above. The overall qualitative agreement between FLH and simulations implies that FLH is a good first order theory to analyse such large quench nonequilibrium simulations within the single phase. One of the reasons for this is that the FLH has built in correctly the infinite-time, small-k structure factor in the theory.

Finally we note that similar results have been observed in the laboratory experiments on binary fluid mixtures[16] and to a limited extent in the computer simulations of three dimensional binary alloys.[7] These are however for states much closer to the critical point. The main theme of our results is that the development of a peak in the structure factor and its subsequent time evolution, analogous to the coarsening phenomena during the spinodal decomposition, is a consequence whenever a large nonequilibrium perturbation of the system takes it to a new final equilibrium state with a larger value of susceptibility (compressibility or $\partial \mu / \partial c$, etc.). And such behaviour of $S(\mathbf{k}, t)$ leads to interesting transient patterns of the order parameter in the configuration space during the nonequilibrium time evolution.

ACKNOWLEDGEMENTS

One of us (RCD) would like to thank (i) Katja Lindenberg and John Wheeler for their hospitality at the Institute for Nonlinear Science and the Department of Chemistry, University of California at San Diego, where this manuscript was completed, and (ii) Ramon Peralta and Carmen Varea for the hospitality at the seventeenth winter meeting in Statistical Physics in Oaxtepec. We would also like to gratefully acknowledge the support from (i) the NSERC of Canada, and (ii) IBM-Canada for use of an IBM3081.

REFERENCES

1. Gunton, J.D., San Miguel, M. and Sahni, P.S., *Phase Transition and Critical Phenomena*, Vol. 8, Ed. C. Domb and J.L. Lebowitz (1983).
2. Heermann, D.W., *Computer Simulation Methods in Theoretical Physics*, Springer (1986).
3. Throughout the text, the word "temperature" is used to indicate the mean kinetic energy per degree of freedom; see also[10] below.
4. Rogers, T.M., Elder, K.R. and Desai, R.C., "A Numerical Study of the Late Stages of Spinodal Decomposition", *Phys. Rev.* **B1**, (in press) (1988); Elder, K.R., Rogers, T.M. and Desai, R.C., *Early Stages of Spinodal Decomposition for the Cahn-Hilliard-Cook Model of Phase Separation*, (preprint) (January 1988).
5. Binder, K., *Condensed Matter Research Using Neutrons*, p. 1, Ed. S.W. Lovesey and R. Scherm, Plenum (1984); Binder, K. and Heermann, D.W., *Scaling Phenomena in Disordered Systems*, p. 207, Ed. R. Pynn and A. Skjeltorp, Plenum (1985).
6. Gunton, J.D. and Droz, M., *Introduction to the Theory of Metastable and Unstable States*, Springer (1983).

7. Koch, S.W., Desai, R.C. and Abraham, F.F., *Phys. Rev.* **A27**, 2152 (1983).
8. Koch, S.W., Desai, R.C. and Abraham, F.F., *Phys. Rev.* **A26**, 1015 (1982).
9. The $S(\mathbf{k}, t)$ described here should not be confused with the *unequal* time correlation function of the order parameter that is observed in the inelastic neutron and light scattering experiments performed on systems in *thermal equilibrium*. For a dense hard sphere fluid in thermal equilibrium, an interesting computer simulation study by Alley, W.E., Alder, B.J. and Yip, S., *Phys. Rev.* **A27**, 3174 (1983) computes and discusses such a correlation function. But it is a different physical quantity for which the same symbol is often used in literature.
10. We use dimensionless units *in the text*: For a 2-d, Lennard-Jones fluid of N particles in area A, $\rho_0 = N\sigma^2/A$, $T_0 = k_B T/\epsilon$, where σ, ϵ are the potential parameters and k_B the Boltzmann's constant. The distances are measured in units of σ and time in units of $\sigma(m/\epsilon)^{1/2}$. For states away from the critical point, the correlation time is of the order of $\sigma(m/\epsilon)^{1/2}$. For values corresponding to argon-argon potential, these units are 3.4×10^{-10} m and 2.16×10^{-12} s respectively. For ease in comparison to other results in literature,[7,8] time is shown in picoseconds in the figures.
11. Gass, D.M., *J. Chem. Phys.* **54**, 1898 (1971).
12. Henderson, D., *Mol. Phys.* **34**, 301 (1977).
13. Baker, J.A., Henderson, D. and Abraham, F.F., *Physica* **106A**, 226 (1981).
14. In the CE simulations, value of the infinite time temperature T_0 depends on (ρ_i, T_i). To obtain the desired value of the final temperature T_0, one has to first determine a state (ρ_i, T_i) with the same total energy and then quench the system to it.
15. Fehder, P.L., *J. Chem. Phys.* **50**, 2617 (1969).
16. Wong, N-C and Knobler, C.M., *Phys. Rev. Lett.* **43**, 1733 (1979); *Phys. Rev. Lett.* **45**, 498 (1980).
17. Marro, J., Bortz, A.B., Kalos, M.H. and Lebowitz, J.L., *Phys. Rev.* **B12**, 2000 (1975) (see Figures 5 and 11).